ENERGY SCIENCE, ENGINEERING AND TECHNOLOGY

U.S. OIL SOURCES

TURMOIL AND TRANQUILITY

ENERGY SCIENCE, ENGINEERING AND TECHNOLOGY

Additional books in this series can be found on Nova's website under the Series tab.

Additional E-books in this series can be found on Nova's website under the E-books tab.

ENERGY SCIENCE, ENGINEERING AND TECHNOLOGY

U.S. OIL SOURCES

TURMOIL AND TRANQUILITY

NATALIE B. KELLEHER
EDITOR

Nova Science Publishers, Inc.
New York

Library of Congress Cataloging-in-Publication Data

ISBN 978-1-62100-415-8

Published by Nova Science Publishers, Inc. † New York

CONTENTS

PREFACE

Despite long standing concern by policy makers, U.S. oil imports have generally increased for decades. This book explores U.S. oil sources with a focus on the U.S.-Canada energy relationship; the Keystone XL Pipeline project; Middle East and North Africa unrest and the implications for oil and natural gas markets; Egypt's turmoil on the global oil and natural gas supply; hedge fund speculation and oil prices and The Strategic Petroleum Reserve and Refined Product Reserves Act.

Chapter 1 - Despite long standing concern by policy makers, U.S. oil imports have generally increased for decades. Two periods stand out as exceptions: the early 1980s and the last five years. Both periods were characterized by high oil prices, economic volatility, and attention to energy policy. U.S. oil imports fell each year between 2005 and 2010 to reach just under 50% of U.S. liquid fuel consumption, its lowest level since 1997. The economic downturn and higher oil prices were a drag on oil consumption, while price-driven private investment and policy helped increase domestic supply of oil and oil alternatives.

Chapter 2 - The United States and Canada, while independent countries, effectively comprise a single integrated market for petroleum and natural gas. Canada is the single largest foreign supplier of petroleum products and natural gas to the United States—and the United States is the dominant consumer of Canada's energy exports. The value of the petroleum and natural gas trade between the two countries totaled nearly $100 billion in 2010, helping to promote general economic growth and directly support thousands of energy industry and related jobs on both sides of the border. Increased energy trade between the United States and Canada—a stable, friendly neighbor—is viewed by many as a major contributor to U.S. energy security. The U.S.-Canada energy relationship is increasingly complex, however, and is undergoing fundamental change, particularly in the petroleum and natural gas sectors.

Chapter 3 - Canadian pipeline company TransCanada has filed an application with the U.S. Department of State to build the Keystone XL pipeline, which would transport crude oil from the oil sands region of Alberta, Canada, to refineries in the United States. Keystone XL would have the capacity to transport 830,000 barrels per day, delivering crude oil to the market hub at Cushing, OK, and further to points in Texas. The project is expected to cost more than $7.0 billion, of which at least $5.4 billion would be spent on the U.S. portion. TransCanada is planning to build a short additional pipeline so that oil from the Bakken formation in Montana and North Dakota can also be carried on the Keystone XL pipeline.

Chapter 4 - Political unrest in the Middle East and North Africa (MENA) has contributed to higher oil prices and added instability to energy markets. Supply disruptions and fears about the possible spread of unrest to major exporters have pushed prices higher. Even if the crisis abates, some risk premium may persist to the degree that market participants fear such an event could occur again.

Chapter 5 - The change in Egypt's government will likely not have a significant direct impact on the global oil and natural gas markets. There may be some short-term movements in price, mostly caused by perceived instability in the marketplace, but these would most likely be temporary. However, prolonged instability that raises the specter of spreading to other oil and natural gas producers in the region would likely add to upward price pressures. Although Egypt is considered an energy producer or net exporter overall, its oil and natural gas exports are not large enough to affect regional or global prices. The most serious impact would be on regional recipients of its natural gas exports.

Chapter 6 - Dramatic swings in crude oil prices have led Congress to examine the functioning of the markets where prices are set. A particular concern is that financial speculators may at times drive prices above the level justified by supply and demand. Most oil speculators do not produce or take delivery of commercial quantities of oil; rather, they trade financial contracts whose value is linked to the price of oil. These derivative contracts—futures, options, and swaps—allow speculators to profit if they can forecast price trends or exploit new arbitrage opportunities. Derivatives also permit oil companies, airlines, utilities, and other energy-consuming or energy-producing firms to reduce or "hedge" price risk by locking in today's price for transactions that will occur in the future. Hedgers and speculators trade on regulated futures exchanges in a continuous auction market. Prices set there serve as benchmarks for many physical oil transactions.

Chapter 7 - Congress authorized the Strategic Petroleum Reserve (SPR) in the Energy Policy and Conservation Act (EPCA, P.L. 94-163) to help prevent a repetition of the economic dislocation caused by the 1973-1974 Arab oil embargo. The Department of Energy (DOE) manages the SPR, which comprises five underground storage facilities, solution-mined from naturally occurring salt domes in Texas and Louisiana. The Energy Policy Act of 2005 (EPAct) authorized SPR expansion to a capacity of 1 billion barrels, but physical expansion of the SPR has not proceeded beyond 727 million barrels — its inventory at the end of 2010. In addition, a Northeast Home Heating Oil Reserve (NHOR) holds 2 million barrels of heating oil in above-ground storage.

In: U.S. Oil Sources
Editor: Natalie B. Kelleher
ISBN: 978-1-62100-415-8

Chapter 1

U.S. OIL IMPORTS: CONTEXT AND CONSIDERATIONS

Neelesh Nerurkar

SUMMARY

Despite long standing concern by policy makers, U.S. oil imports have generally increased for decades. Two periods stand out as exceptions: the early 1980s and the last five years. Both periods were characterized by high oil prices, economic volatility, and attention to energy policy. U.S. oil imports fell each year between 2005 and 2010 to reach just under 50% of U.S. liquid fuel consumption, its lowest level since 1997. The economic downturn and higher oil prices were a drag on oil consumption, while price-driven private investment and policy helped increase domestic supply of oil and oil alternatives.

Among the sources of net U.S imports, about a quarter of net imports come from Canada and another half come from countries that are members of the Organization of the Petroleum Exporting Countries (OPEC). Almost 20% of imports come from the Persian Gulf (from OPEC countries in the region). Global oil supply disruptions can shift import trends and raise prices for oil produced at home and imported. This is true even if the disruption occurs in countries that are not normally sources of U.S. imports. Even anticipation of disruption risks can have similar impacts.

Reliance on oil imports is of particular concern to policy makers when oil prices increase. Increased fuel costs for households and businesses can reduce spending on other goods and services that might be domestically produced, send more wealth abroad, and cause economic dislocations, such as greater unemployment. Economic forecasters estimate a sustained $10 increase in the per barrel price of oil can reduce U.S. economic growth by 0.2% in part due to how much U.S. consumption is met by imports. However, even in a scenario where the U.S. produced as much oil as it consumed, an increase in international oil prices would still raise oil costs for households and businesses and cause economic dislocations, but wealth associated with the increase may remain in the country.

Oil import volumes are expected to remain roughly flat over the next two decades. There is a growing consensus that U.S. imports passed their high-watermark in 2005. Such forecasts are predicated on expectations that current laws supporting fuel efficiency and domestic supply are not reversed and that oil prices continue to rise. While volumes may remain flat, rising prices could still increase the cost of oil imports.

There is congressional interest in further reducing the potential risks posed by import dependence. Policy options generally fall into four categories:

- Direct trade policy regarding oil imports,
- Long-term measures aimed at reducing the need for imports through greater domestic supply (conventional and alternative) and greater fuel efficiency to reduce demand,
- Short term energy policy tools like the release of oil stored in the Strategic Petroleum Reserve (SPR), and
- Foreign policy measures aimed at making foreign sources of oil more secure (not covered in this report).

INTRODUCTION

Oil is a critical resource for the U.S. economy. It meets nearly 40% of total U.S. energy requirements, including 94% of the energy used in transportation and 41% of the energy used by the industrial sector. Unlike other forms of energy such as coal and natural gas, which are largely supplied from domestic sources, half of U.S. oil consumption is currently supplied from foreign sources.

The United States has been concerned about dependence on foreign oil since it became a net oil importer in the late 1940s. Those concerns grew with import levels, especially in periods of high or rising oil prices. Imports have generally increased over the last six decades, except for a period following the oil spikes of the 1970s and again in the last five years. See **Figure 1**. Oil import volumes peaked in 2005 and then declined through 2010 as a result of economic and policy-driven changes in domestic supply and demand. However, oil import costs have increased due to rising prices, which more than offset the savings from lower import volumes.

Interest in oil imports has climbed again in recent months as oil prices rebounded to over $100 per barrel. President Obama highlighted the Administration's concerns around recent energy developments in a speech on March 30, 2011, proposing policies to cut oil imports by one-third. Congress has also voiced concern and introduced proposals that aim to reduce the risks posed by reliance on oil imports. This report first explains recent oil import trends. It then discusses long- and short-term policy considerations.

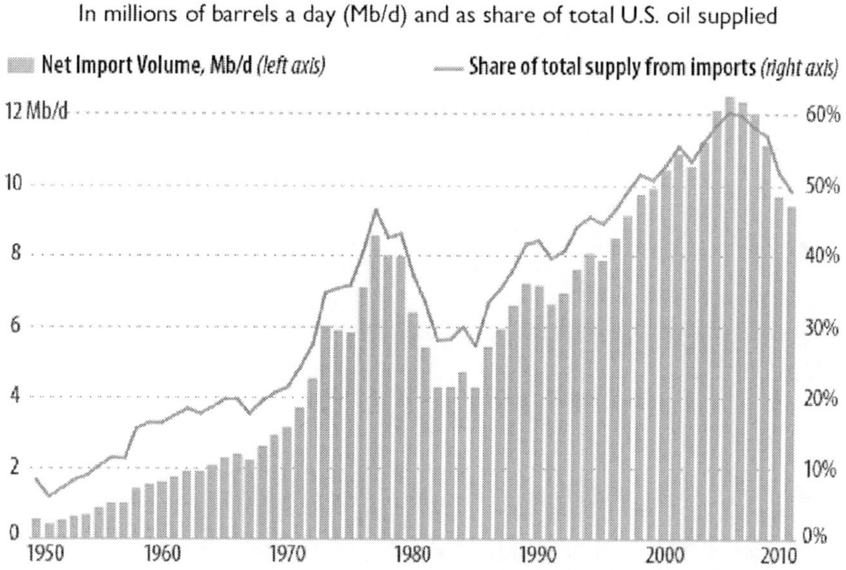

Data Source: Energy Information Administration, U.S. Department of Energy, *Short-Term Energy Outlook (STEO)*, March 8, 2011, *http://www.eia.doe.gov/emeu* /steo/pub/contents.html. Also Energy Information Administration U.S. Department of Energy, *Annual Energy Review (AER)*, August 19, 2010, http://www.eia.doe.gov/emeu/aer/contents.html.

Figure 1. Net Oil Imports.

U.S. OIL IMPORTS

Oil Import Fell between 2005 and 2010

U.S. net oil imports, including crude oil, refined petroleum products, and other liquid fuels,[1] fell each year between 2005 and 2010, from 12.4 million barrels a day (Mb/d) in 2005 to 9.4 Mb/d in 2010.[2] Only 49% of U.S. oil consumption in 2010, which totaled 19.2 Mb/d, was met with imported oil, down from 60% in 2005. Net imports are at their lowest level since 1997, both in terms of volumes of oil imported and percentage of consumption, see **Figure 1**. This decline has resulted from lower consumption and increased domestic oil production.

Total U.S. consumption of oil fell from 2005 to 2009. Despite a small increase in 2010, it remained about 1.7 Mb/d below 2005 levels. The decline is due largely to the recession, which started in December 2007, as well as greater efficiency in reaction to several years of high oil prices.

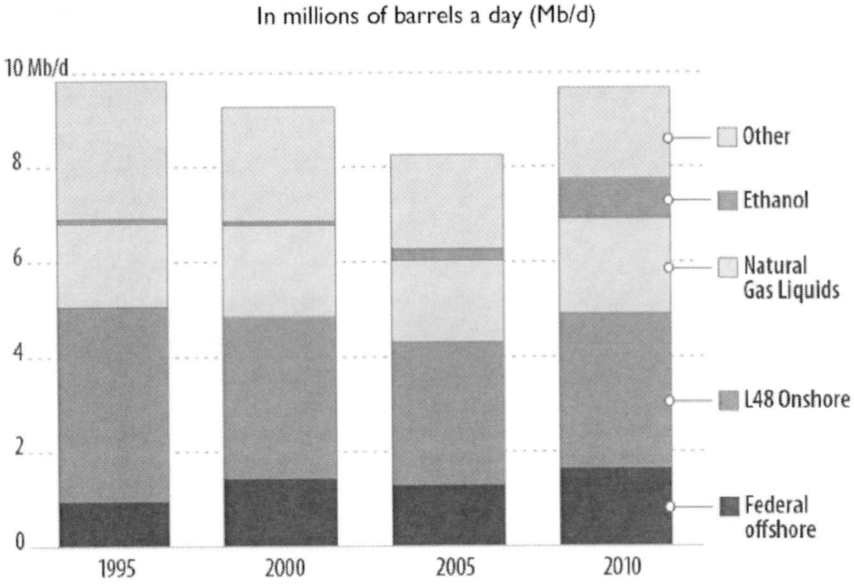

Data Source: EIA, *STEO*.

Notes: Other includes production from Alaska, refinery processing gain, renewable fuels and biofuels and oxygenates excluding fuel ethanol. L48 refers to the contiguous 48 states.

Figure 2. Domestic Oil and Other Liquid Fuels Supply.

Table 1. Petroleum Trade Balance by Liquid Fuel Type in million barrels a day

	2005	2006	2007	2008	2009	2010		
	Balance	Balance	Balance	Balance	Balance	Imports	Exports	Balance
Total	**(12.5)**	**(12.4)**	**(12.0)**	**(11.1)**	**(9.7)**	**(11.8)**	**2.3**	**(9.4)**
Unfinished liquid fuels	(11.5)	(11.7)	(11.6)	(11.5)	(10.4)	(10.7)	0.3	(10.4)
Crude Oil	(10.1)	(10.1)	(10.0)	(9.8)	(9.0)	(9.2)	0.0	(9.1)
Other unfinished	(1.4)	(1.6)	(1.6)	(1.7)	(1.4)	(1.5)	0.2	(1.3)
Finished Refined Products	(1.1)	(0.7)	(0.4)	0.3	0.7	(1.1)	2.0	1.0
Gasoline	(0.5)	(0.3)	(0.3)	(0.1)	(0.0)	(0.1)	0.3	0.2
Distillate	(0.2)	(0.2)	(0.0)	0.3	0.4	(0.2)	0.7	0.4
Residual Fuel Oil	(0.3)	(0.1)	(0.0)	0.0	0.1	(0.4)	0.4	0.0
Petrochemical Feedstock	(0.3)	(0.3)	(0.2)	(0.2)	(0.1)	(0.1)	0.0	(0.1)
Petroleum Coke	0.3	0.3	0.3	0.4	0.4	(0.0)	0.4	0.4

Source: EIA, *Petroleum and Other Liquids: Imports/Exports & Movements*, http://www.eia.doe.gov/petroleum/data.cfm.

Notes: 2005-2009 figures are exports minus imports. Parenthesis indicate a negative balance (more imports than exports). 2010 figures show imports, exports, and balance. "Other unfinished" includes non-crude unfinished liquid fuels, such as NGLs, gasoline blending components, and heavy gas oils.

Meanwhile, higher prices also helped support greater domestic production. Production had been falling steadily since 1985 to 8.3 Mb/d in 2005. Since then, increased supplies from ethanol, offshore oil production, onshore production in the lower 48 states, and natural gas liquids raised domestic output by 1.5 Mb/d. The decline in consumption and increase in domestic supply led to net imports falling by 3.1 Mb/d or 25% between 2005 and 2010. See **Figure 2**.

Net Versus Gross: Total U.S. Imports and Exports

Net imports are gross imports minus exports. In 2010, U.S. gross oil imports were about 11.8 Mb/d, down from a peak of 13.7 Mb/d in 2005. Meanwhile, exports of oil increased from 1.2 Mb/d to 2.3 Mb/d, nearly all of which is in the form of refined petroleum products. Exports to Mexico, Europe, and Asia have grown. These shifts reflect the domestic balance of supply and demand discussed above—consumption fell and production increased. See details about the differences between volume and sources of net and gross imports in **Appendix A** and **Appendix B**. Note, both net and gross measures are relevant to energy security concerns. Net measures can be more relevant to economic concerns around oil imports and are the focus of this report.

U.S. imports are primarily crude oil, while exports are almost entirely refined petroleum products (see **Table 1**). Traditionally, much of U.S. oil exports are heavier products from refineries, such as petroleum coke, for which there is limited demand in the United States, or products that do not meet U.S environmental standards.[3] Although the United States is a net oil importer, it does have significant refining capacity which has allowed it to export some finished refined products. Moderating demand at home and rising demand abroad led to U.S. exports of distillate since 2008 and of finished motor gasoline in 2010. Market conditions—the particular needs (and ability to pay) of specific buyers and sellers in different parts of the world at a given time—can drive the export of some petroleum product cargos even when the United States generally imports oil.

> Unlike many nations around the world in which oil production, refining, trade, and marketing are dominated by government-owned national oil companies, these functions are carried out by numerous privately owned companies in the United States. The size and characteristics of U.S. imports and their variation over time result from myriad decisions by a multitude of individual U.S. oil companies in response to domestic and global market opportunities and demands. Although these companies operate under U.S. national policies, and their decisions are affected by these policies, there is no centralized control of oil imports and exports in the United States.

Of the petroleum products that the United States exports, half goes to other countries in the Western Hemisphere. In 2010, nearly 20% of exports went to Mexico and nearly 10% to Canada (both send substantially more oil to the United States than they import from the United States). Brazil, Chile, Panama, and Ecuador each received 3%-5% of exports. About 1% went to Venezuela.

Sources of Oil Imports

Most U.S. oil imports come from Canada, Mexico, and members of the Organization of the Petroleum Exporting Countries (OPEC).[4] In 2010, about 25% of U.S. consumption (50% of net imports) was supplied by OPEC countries: 9% from Persian Gulf OPEC members, the rest from Latin American and Africa members of the cartel.[5] Net imports from Canada and Mexico met another 16% of U.S. consumption (33% of net imports); see **Figure 3**. A list of the top 10 sources of U.S. net oil imports is available in **Appendix A**. The largest volume of imports from any single country came from Canada —2.3 Mb/d, more than twice the amount of the next largest source (Saudi Arabia, on a net basis).

Oil trading patterns can shift due to market conditions, but much of the oil that companies import to the United States comes from the same countries month to month. While many kinds of crude oil are interchangeable to some degree—which reinforces the globally integrated nature of the oil market—each has unique characteristics. Refineries optimize their operations for particular kinds of crude.[6] Proximity, infrastructure, and trading relationships may also contribute to normal trade patterns and determine which crude oils are consumed in certain countries.

Over the last several years, imports of oil from many countries fell with the general trend of falling imports. The decline in imports from some specific countries was exacerbated by their declining output. For example, declining oil output from Mexico, the United Kingdom, and Norway contributed to lower imports from those countries.[7] In addition, imports from OPEC also fell in part because OPEC countries intentionally curtailed their output in 2008-2009 to stem the decline in oil prices; see **Figure 4**.[8]

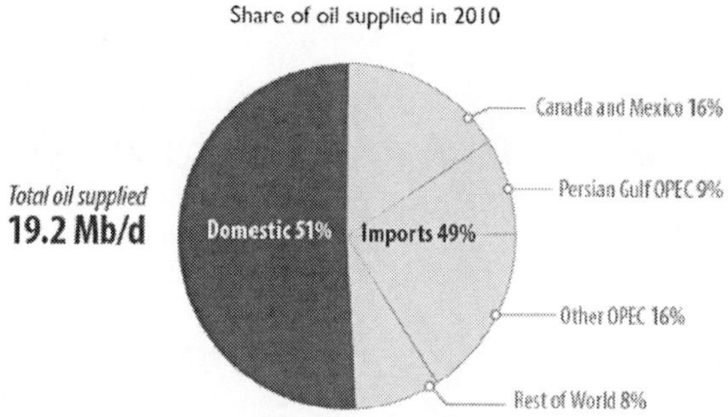

Source: EIA, *STEO*, and EIA, "U.S. Net Imports by Country," Feb 2, 2011, http://www.eia.doe.gov/dnav/pet/pet_move_neti_a_EP00_IMN_mbblpd_m.htm.

Notes: "Persian Gulf OPEC" members includes Saudi Arabia, Kuwait, United Arab Emirates, Iraq and Qatar (the United States does not import oil from Iran). "Other OPEC" includes Angola, Nigeria, Algeria, Libya, Venezuela, and Ecuador. For more detail, see **Appendix A**.

Figure 3. Net Imports as a Share of Total Oil Supplied: Key Foreign Sources.

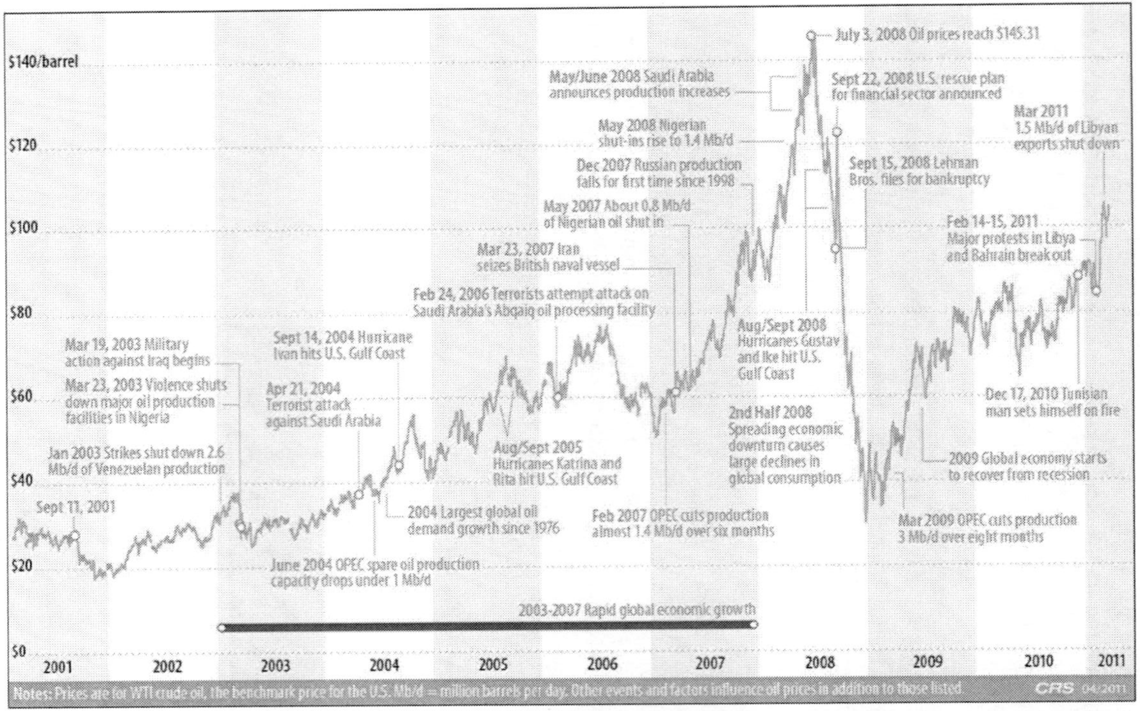

Source: CRS. Includes data from EIA oil price databases (http://www.eia.doe.gov/dnav/pet/pet_pri_spt_s1_d.htm) and Annual Oil Market Chronology (http://www.eia.doe.gov/emeu/cabs/AOMC/Overview.html), and IMF World Economic Outlook databases (http://www.imf.org/external/ns/cs.aspx?id=28).

Figure 4. Selected Events That Affected the Price of Oil.

Imports from Canada are a notable exception to the general downward trend. Imports from Canada have increased by roughly 0.3 Mb/d since 2005 due to rising production of oil sands and because oil pipelines tie Canadian exports to U.S. markets.[9] Imports have also increased from Colombia, Brazil, and Azerbaijan—all countries that experienced significant increases in national oil production. Trade with Colombia and Brazil benefits from proximity of the U.S. market to Latin America.

Supply Disruptions Can Shift Imports and Raise Prices

Shifting trade flows arising from abrupt supply disruptions are a source of particular concern for the oil industry and policy makers. Sudden shifts in supply can require rapid adjustment by industry and raise global oil prices. When strikes reduced oil output from Venezuela, traditionally a large supplier of oil to the United States, from 3.3 Mb/d in November 2002 to 0.7 Mb/d in January 2003, U.S. imports from Venezuela fell by more than 1 Mb/d. Other examples include when periodic militant attacks shut down oil production facilities in Nigeria and when summer/fall hurricanes shut down Mexico's offshore production; both are major sources of U.S. imports. Such events may lead refiners to draw on oil inventories and/or bid for supply from elsewhere in the world.

Domestic supply disruptions can also shift trade flows. After hurricanes Katrina and Rita shut in oil production in the U.S. Gulf of Mexico, U.S. imports increased by around 0.7 Mb/d between July and October 2005. The increase was in refined products; hurricanes shut down more refining capacity than crude oil production. Crude imports fell.

Supply disruption in countries that are not traditionally major sources of U.S. imports may still have significant implications for the United States because they raise the price of oil worldwide. The oil market is globally integrated, refiners can shift the crude they use, and refined products are interchangeable commodities; so a disruption anywhere can affect oil prices everywhere. For instance, the United States imported only around 0.1 Mb/d of oil from Libya in 2010. (For context, the U.S. consumed about 19.2 Mb/d in 2010.) Most of Libya's crude supply went to Europe. But when unrest shut down Libya's exports in February 2011, global prices rose, including prices for oil imported into the United States from elsewhere and oil produced domestically. Global supply was reduced and European refiners had to look to other oil sources, bidding up those oil prices to secure substitute supplies.[10]

The price of oil may rise until it makes up for the amount of supply no longer available due to the disruption. This can occur by price rising enough that some consumers no longer demand oil and/or suppliers bring additional production to market.[11] Many oil producers and consumers are inelastic to price changes when considering how much to supply or consume, especially in the short run, so seemingly small disruptions can lead to more significant percent changes in the price of oil.

Even anticipation of disruptions can contribute to higher oil prices. Buyers and sellers of oil make risk-weighted decisions now about future commercial and financial needs. Anticipated disruption risks affect the price at which they are willing to buy and sell oil. Arguably, a significant portion of the increase in oil prices from unrest in Libya, Egypt, Bahrain, and elsewhere is attributable to concerns that unrest could spread to other oil exporters in the Middle East and North Africa.

Disruptions to oil production reduced supply, slowed supply growth in recent years, and created concerns about future supply. This combined with rising oil demand, resulting from rapid economic growth in several countries, as well as other financial, geologic, commercial, and political factors, contributed to the rise in oil prices during the 2000s. Some selected events that played a role in recent price developments are presented in **Figure 4**.

Oil Imports and the Trade Deficit

The aggregate national cost of oil imports is a function of the volume of oil imported and the price of that oil. The United States spent about $265 billion on net oil imports in 2010. Import volumes and resulting costs fell with the economic downturn. Lower oil prices also helped push down oil import costs in 2009. Import costs increased again with higher oil prices in 2010, as seen in **Table 2**. Some estimates suggest import costs could be $100 billion dollars higher in 2011 as prices have risen from 2010 levels.[12] This would surpass the previous peak cost of $386 billion reached in 2008, the year with the highest average annual oil price in history to date: $99.67 per barrel for WTI.

Despite net oil import volume falling since 2005, net import costs have generally increased. The increase in oil prices has more than offset the decline in import volumes. While 25% fewer barrels were imported in 2010 versus 2005, the price U.S. oil refiners paid to import barrels of crude averaged 55%

higher (natural gas liquids prices made similar increases).[13] This price increase has caused oil imports to constitute a larger share of the trade deficit, see **Figure 5**.

Table 2. Energy in the Trade Balance in millions of dollars

	Imports		Exports		Trade Balance	
	2009	2010	2009	2010	2009	2010
Total oil	253,689	335,881	49,176	70,764	(204,513)	(265,117)
Crude Oil	188,712	252,064	972	1,328	(187,740)	(250,736)
Other Oil	64,977	83,817	48,204	69,436	(16,773)	(14,381)
Natural Gas	16,056	16,109	3,286	4,488	(12,770)	(11,621)
Electricity	2,075	2,071	562	648	(1,513)	(1,423)
Nuclear fuel	5,363	5,641	2,283	1,896	(3,080)	(3,745)
Coal	2,160	2,136	6,521	10,463	4,361	8,327
Total Energy	**279,343**	**361,838**	**61,828**	**88,259**	**(217,515)**	**(273,579)**
Total Goods and Services	**1,945,705**	**2,329,894**	**1,570,797**	**1,834,166**	**(374,908)**	**(495,728)**
Energy Share	*14%*	*16%*	*4%*	*5%*	*58%*	*55%*
Oil Share	*13%*	*14%*	*3%*	*4%*	*55%*	*53%*

Source: U.S. Census Bureau, "U.S. International Trade in Goods and Services Report," March 10, 2011, http://www.census.gov/foreign-trade/data/index.html. Figures on a balance of payments basis.

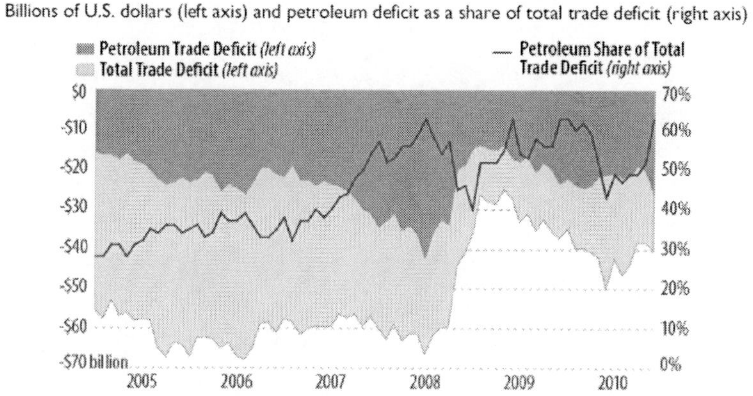

Data source: U.S. Census Bureau, "U.S. International Trade in Goods and Services Report," March 10, 2011, http://www.census.gov/foreign-trade/data/index.html. Figures on a balance of payments basis.

Figure 5. Petroleum in the Monthly Trade Deficit.

Being a net importer of a particular good is not necessarily negative for an economy; the United States imports many products because they would be more costly to consumers, businesses, or taxpayers were they produced domestically. But import dependence for a good can also have negative impacts, particularly if it contributes to long-term trade deficits. Trade deficits are ultimately financed by borrowing from abroad, which generates financial obligations for future repayment.

Greater national oil import dependence can also amplify the negative economic impacts of oil price increases. An increase in the price of crude oil can quickly translate to higher prices for refined oil products like gasoline. Rising oil product prices strain the budgets of households or businesses, reducing their savings and/or spending on other goods and services, some of which would have been produced domestically. For imported oil, that wealth is sent abroad, and only a portion of it is returned via now wealthier oil exporting countries buying more U.S. products. Even with oil from domestic sources, an oil price increase redistributes wealth within the economy. This may cause temporary dislocation as businesses and workers adjust to the change. The impact can be exacerbated when the price change is faster as it makes the dislocations more abrupt.

In millions of barrels a day

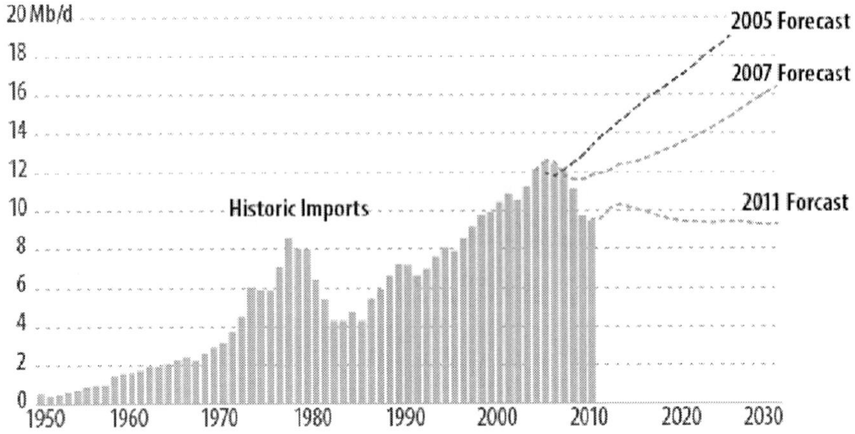

Data source: EIA, *AEOs* from 2005, 2007, and 2011. http://www.eia.doe.gov/forecasts /aeo/index.cfm.

Figure 6. Lower Expectations for Future Oil Imports.

These and other economic impacts of rising oil prices and import costs are varied and complex. Economic analysts estimates that the impact of a sustained $10 per barrel increase in the price of oil could result in about 0.2% lower economic growth and 120,000 fewer jobs in the first year after the increase.[14] Rapidly rising oil prices likely contributed to the U.S. economic recession in 2007-2008.[15] Some suggest that every major oil price increase since the 1970s has been associated with a recession.[16]

If the United States were not a net importer of oil—if it produced as much oil as it consumed—rising oil prices would not increase the import bill, but would still negatively impact the budgets of many U.S. households and businesses. Wealth would be redistributed from oil consumers to oil producers within the economy. Oil prices paid by U.S. consumers for petroleum products would still be affected by international events as long as oil trade was permitted. Other impacts of higher oil prices, like inflation and unemployment, may continue to be economic concerns.

Looking Forward: Import Dependence Passed Peak?

A consensus is generally emerging among energy analysts that U.S. oil imports may be past their peak, reached in 2005, and are likely to remain roughly flat over the next 20 years.[17] This conclusion has been reached as the forces that pushed imports down in recent years have become evident. While some of these recent forces are temporary—for example, those related specifically to the recession—some are expected to persist: high oil prices, rising efficiency, and greater domestic production of natural gas liquids, biofuels, and crude oil. See the evolution over time of EIA projections of future oil imports in **Figure 6**.

POLICY CONSIDERATIONS

Oil import dependence has been a policy concern since the United States became a net oil importer in the late 1940s. Nonetheless, imports generally increased since then, except during the early 1980s and in the late 2000s, as shown in **Figure 1**. Both periods of falling imports resulted from economic and policy reactions to high oil prices. These in turn increased domestic liquid fuels production and reduced demand, bringing down imports.[18] After oil prices collapsed in 1985, economic drivers waned, policy initiatives were rolled back, and imports climbed once more. Today, forecasters expect that some of these drivers are more durable—largely because prices are expected to remain high— and that imports may be past their peak levels. But there remains policy interest in reducing oil imports further. Such concerns were underscored by recent unrest in the Middle East and North Africa, giving rise to calls for a range of policies to reduce import related risks.

Policy responses to address risks posed by oil imports tend to fall into one of four categories:

- *Trade policy*: Policies that directly address the import of oil, such as tariffs, quotas, or sanctions.
- *Reduce the need for imports*: Most legislative initiatives to reduce oil imports are indirect — they target raising domestic production of oil and oil alternatives or reduction of oil demand. They work through creating new incentives and greater policy intervention, or by removing existing policy hurdles to private investment.
- *Short-term policy*: Amending rules around the government's Strategic Petroleum Reserve can affect how well prepared the United States is to address oil supply disruptions.
- *Foreign policy*: U.S. diplomatic and military measures can affect the stability of foreign oil sources, though such policies fall beyond the scope of this report.[19]

Direct Policies on Oil Imports

Shortly after the United States became a net oil importer, President Truman appointed the Paley Commission to examine the security of supply for oil and other basic materials. It supported policies for greater domestic production, particularly from oil shale, and efficiency, but also advised against raising trade barriers to imports.[20] Imports continued to rise despite persistent concerns around import dependence. Under powers granted in the Trade Agreements Extension Act of 1958, President Eisenhower imposed quotas that limited imports east of the Rockies to a certain percentage of total consumption.[21] The quota program broke down by 1969 due to rising oil prices and quota loopholes. President Nixon replaced the program with import fee in 1974, set at only a few cents per barrel.[22]

Current tariffs on oil imports range from 5.25¢ to 52.5¢ per barrel depending on the type of petroleum.[23] See various rates in **Appendix C**. Oil and petroleum products from certain countries are subject to duty free treatment under several trade agreements and preferential trade programs enacted by Congress. The North American Free Trade Agreement (NAFTA), the Generalized System of Trade Preferences (GSP), and the African Growth and Opportunities Act (AGOA) account for most of the foregone revenue from waived tariff.[24] At 2010 import levels, these and other waiver programs accounted for about $180 million dollars in foregone revenue in 2010, down from $215 million in 2005 when import volumes were higher.[25]

In the past, an increase in the tariff on oil has been considered as a means to provide an advantage to domestic oil producers and reduce imports. Because tariffs on refined oil products are already at the highest levels permitted by U.S. agreements with the World Trade Organization (WTO), any increase may be limited to crude oil tariffs.[26] The United States has not made any tariff commitments bounding duties on crude oil.[27] But like crude oil, refined products are globally traded. There is a risk that an increase in the tariff on crude oil alone may create a disadvantage for domestic refiners vis-à-vis their foreign competition. Such a tariff could shift imports in the form of crude oil to imports in the form of oil products refined abroad. On the other hand, the availability of foreign refined products imports may also limit how much of an increase in a crude oil tariff could be passed on to domestic consumers.

Restrictions on Oil Exports

The wisdom of allowing exports of oil and petroleum products is sometimes questioned. Export of Alaskan crude oil had been prohibited by the Trans-Alaskan Pipeline Authorization Act of 1973. Congress removed the restriction in 1995. In general, because the market for oil is global, a prohibition on U.S. oil exports is unlikely to contribute to lower prices for U.S consumers. Such a restriction is likely to only cause a rebalancing in the movement of petroleum. Restrictions on exports might, in fact, create inefficiencies in the movement of world oil supplies that could foster less optimal distribution of oil and possibly lead to higher prices in some markets.[28] An alternative way to consider the issue is that exports of oil products less in demand in the United States help offset the import cost of crude oil and other oil products for which there may be greater U.S. demand.

Reducing the Need for Imports: Supply and Demand Policies

President Obama highlighted the long-term nature of potential solutions to oil import challenges in his March 30, 2011 speech on energy. The Administration has targeted reducing oil imports by one-third over the next decade through encouraging greater conventional crude oil production, increased production of biofuels, and efficiency or alternative fuel vehicles.[29] Members of Congress have also introduced a range of proposals aimed at increasing domestic liquid fuels supply or reducing liquid fuels demand. These include creation of new policies or programs, or the removal of existing policy barriers to supply enhancing (or demand dampening) investment. Either sets of options come with associated costs and benefits to be weighed.

Production and use of ethanol as a motor fuel is an example. Fuel ethanol production has increased by roughly 0.6 Mb/d since 2005, when the renewable fuels standard and ethanol excise tax credit were passed.[30] This is the largest single component of the domestic supply growth between 2005 and 2010 discussed above (in volumetric terms). But some critics argue that increased use of corn for biofuels is driving up food prices. Other critics have cited the high cost to tax payers in the form of forgone revenues. The ethanol tax credit reduced federal excise tax revenue by roughly $6 billion in 2009; it cost taxpayers $1.78 for each gallon of gasoline consumption displaced by corn-based ethanol according to Congressional Budget Office estimates.[31] And there are limits to how much fuel ethanol can be absorbed by the U.S. market.[32]

Offshore oil and gas development is another example of tradeoffs being debated. Oil output from the federal offshore region of the Gulf of Mexico was the second largest component of recent domestic supply growth. Output from the region has increased by roughly 0.4 Mb/d since 2005. But rising activity in progressively deeper and riskier waters, coupled with alleged industry and government complacency on safety, created environmental risks that were realized with the Deepwater Horizon oil spill.[33] Risks to human safety and the environment may be reduced with improved regulatory enforcement or new regulatory measures, but the risks may not be eliminated.

There are also demand side examples of policy trade-offs. Congress increased the Corporate Average Fuel Economy standard (CAFE) in 2007, its first increase for cars in two decades. CAFE's impacts can take years to manifest because it takes years to turn over the vehicle fleet. To spur growth in the auto industry and accelerate vehicle efficiency, Congress passed the Car Allowance Rebate System Act of 2009 (CARS, also known as "Cash-for-

Clunkers") which provided a $3,500 or $4,500 rebate for trading in a used car for a more fuel efficient vehicle (amount depends on the fuel economy improvement). The CARS program provided a temporary stimulus that contributed to economic recovery from recession,[34] created jobs,[35] and likely reduced some fuel consumption faster than CAFE. But there were tradeoffs: Accelerating fuel efficiency through CARS may have cost the government significantly more than the estimated future fuel savings for participating consumers.[36] It may only reduce 2020 oil consumption and greenhouse gas emissions by an estimated 0.02%.[37] It accelerated some sales that would have taken place in future months anyways.[38] And by scrapping cars traded in, it may have increased the cost of vehicles for those who purchase used cars in the future.

Strategic Petroleum Reserve

Congress created the Strategic Petroleum Reserve (SPR) in the Energy Policy and Conservation act of 1975 to guard against "severe energy supply interruptions." The policy had been considered since at least the Paley Commission report and was ultimately established in reaction to the 1973-1974 Arab Oil Embargo. The government-owned SPR is now filled to its 727 million barrel of crude capacity, equivalent to roughly 77 days worth of imports at 2010 import levels.[39] The crude oil is stored in five underground salt domes in Texas and Louisiana. The government also holds 2 million barrels of heating oil in above-ground storage in the Northeast Home Heating Oil Reserve (NHOR).[40] The President has the authority to release oil from the SPR.

Most policy options available to address oil import related concerns are long-term in nature. Changes in the energy sector have long lead times and it can take many years for policies affecting supply or demand to have material impacts. The SPR provides a short term tool to respond to sudden supply disruptions at home or abroad. But what constitutes a significant enough supply disruption, or whether the SPR should be used to reduce oil prices, is debated. SPR releases were authorized in 1990-1991 around Iraq's invasion of Kuwait and Operation Desert Storm and also after Hurricane Katrina and Rita.

SPR releases can be coordinated with the International Energy Agency (IEA), whose members have also committed to hold strategic reserves.[41] More than 4 billion barrels of oil are held in stocks of IEA members, of which 1.6 billion barrels are held by member governments. Another 2.6 million barrels are held by companies in IEA countries.[42] European IEA members meet most of their commitment to hold strategic reserves by mandating that industry hold stocks for emergencies. Most of Europe's stock holdings are in the form of refined products, in contrast to strategic stocks held by IEA members in North America and the Pacific, which are largely in the form of crude oil and held by governments. IEA stocks could be brought to market at a maximum rate of 14.4 Mb/d in the first month of an IEA collective action.[43] The IEA has brought oil from strategic stocks to market in a coordinated release on two occasions (never at full capacity): 1991 Gulf War and after hurricanes hit the U.S. Gulf of Mexico in 2005.

As more of the world's oil is being consumed by developing countries, some of those nations have started to develop strategic stocks. Both China and India are building strategic stockpiles of oil. The IEA has engaged with them to explore opportunities for coordination should a strategic release become necessary.

APPENDIX A. NET U.S. OIL IMPORTS

Top 10 Sources of Net Imports in Million Barrels a Day

	2005	**2006**	**2007**	**2008**	**2009**	**2010**
Total	12.5	12.4	12.0	11.1	9.7	9.4
Canada	2.0	2.2	2.3	2.2	2.3	2.3
Saudi Arabia	1.5	1.5	1.5	1.5	1.0	1.1
Nigeria	1.2	1.1	1.1	1.0	0.8	1.0
Venezuela	1.5	1.4	1.3	1.2	1.0	1.0
Mexico	1.4	1.5	1.3	1.0	0.9	0.8
Russia	0.4	0.4	0.4	0.5	0.6	0.6
Algeria	0.5	0.7	0.7	0.5	0.5	0.5
Iraq	0.5	0.6	0.5	0.6	0.5	0.4
Angola	0.5	0.5	0.5	0.5	0.5	0.4
Colombia	0.2	0.1	0.1	0.2	0.2	0.3
All others	2.9	2.5	2.3	1.9	1.5	1.0
Selected Groups and Sub-Groups						
OPEC	5.6	5.5	5.9	5.9	4.7	4.8
• Persian Gulf OPEC Countries	2.3	2.2	2.2	2.4	1.7	1.7

	2005	2006	2007	2008	2009	2010
Non-OPEC	7.0	6.9	6.1	5.2	5.0	4.7
• Canada and Mexico	3.4	3.6	3.5	3.2	3.1	3.2

Source: EIA, http://www.eia.gov/dnav/pet/pet_move_neti_a_ep00_ IMN_mbblpd _a.htm.

Notes: Countries in italics are members of OPEC. Virtually all oil trade with the Persian Gulf is with OPEC members in the region.

APPENDIX B. GROSS U.S. OIL IMPORTS

Top 10 Sources of Gross Imports in Million Barrels a Day

	2005	2006	2007	2008	2009	2010
Total	13.7	13.7	13.5	12.9	11.7	11.8
Canada	2.2	2.4	2.5	2.5	2.5	2.5
Saudi Arabia	1.7	1.7	1.5	1.3	1.2	1.3
Nigeria	1.5	1.5	1.5	1.5	1.0	1.1
Venezuela	1.2	1.1	1.1	1.0	0.8	1.0
Mexico	1.5	1.4	1.4	1.2	1.1	1.0
Russia	0.4	0.4	0.4	0.5	0.6	0.6
Algeria	0.5	0.7	0.7	0.5	0.5	0.5
Iraq	0.5	0.6	0.5	0.6	0.5	0.4
Angola	0.5	0.5	0.5	0.5	0.5	0.4
Colombia	0.2	0.2	0.2	0.2	0.3	0.4
All others	3.6	3.4	3.3	3.1	2.9	2.5
Selected Groups and Sub-Groups						
OPEC	5.6	5.5	6.0	6.0	4.8	4.9
• Persian Gulf OPEC Countries	2.3	2.2	2.2	2.4	1.7	1.7
Non-OPEC	8.1	8.2	7.5	7.0	6.9	6.9
• Canada and Mexico	3.8	4.1	4.0	3.8	3.7	3.8

Source: EIA, http://www.eia.gov/dnav/pet/pet_move_impcus_a2_nus_ep00_im0_mbbl_m.htm.

Notes: Countries in italics are members of OPEC. Virtually all oil trade with the Persian Gulf is with OPEC members in the region.

Appendix C. Petroleum Tariff Rates

Normal Petroleum Import Tariffs Under the Harmonized Tariff Schedule of the United States (HTSUS)

Petroleum Categories	Normal Tariff
Petroleum oils and oils from bituminous minerals, crude, testing 25 degrees API gravity or more.	10.5¢/bbl
Petroleum oils and oils from bituminous minerals, crude, testing under 25 degrees API gravity.	5.25¢/bbl
Naphthas (excluding motor fuel/motor fuel blend stock) from petroleum oils and bitumen minerals (other than crude) or preparations 70% plus by weight from petroleum oils.	10.5¢/bbl
Distillate and residual fuel oil (including blends) derived from petroleum or oils from bituminous minerals, testing under 25 degrees API gravity.	5.25¢/bbl
Light oil mixture of hydrocarbons from petroleum oils and bitumen minerals (other than crude) or preparations 70% plus by weight from petroleum oils, not otherwise specified, not over 50% any single hydrocarbon.	10.5¢/bbl
Light oil motor fuel from petroleum oils and bituminous minerals (other than crude) or preparations 70% plus by weight from petroleum oils.	52.5¢/bbl
Distillate and residual fuel oil (including blends) derived from petroleum oils or oil of bituminous minerals, testing 25 degree API gravity.	10.5¢/bbl
Lubricating oils, with or without additives from petroleum oils and bitumen minerals (other than crude) or preparations 70% plus by weight from petroleum oils.	84¢/bbl
Light oil motor fuel blending stock from petroleum oils and bituminous minerals (other than crude) or preparations 70% plus by weight from petroleum oils.	52.5¢/bbl

Source: Harmonized Tariff Schedule Chapter 27. Categories identified by the International Trade Administration.

Notes: "bbl" stands for barrel (42 gallons). API gravity is a measure for the specific gravity of oil developed by the American Petroleum Institute and others. The higher the number, the lighter and less dense the oil. Lighter oils tend to require less processing and yield more valuable products.

End Notes

[1] As is convention in oil market analysis, this report will reference the term "oil" to include crude oil, refined petroleum products, and other liquid fuels, including substitutes such as natural gas liquids and biofuels. Natural gas liquids (NGLs) are liquid fuels that can be extracted from natural gas. Note, NGLs are different from liquefied natural gas (LNG), which is a means of transporting natural gas. "Net" oil imports means total oil imports less oil exports.

[2] Energy Information Administration (EIA), U.S. Department of Energy, Short-Term Energy Outlook (STEO), March 8, 2011. Unless otherwise noted, all data in this report comes from EIA's STEO, the *Annual Energy Outlook* (AEO), *Petroleum & Other Liquids*, or *International Energy Statistics*. www.eia.doe.gov.

[3] CRS Report R40120, *U.S. Oil Exports*, by Robert Bamberger.

[4] OPEC is a cartel of major oil exporting countries that attempts to adjust global oil supply to manage oil prices.

[5] Note, the United States does not import any oil from Iran.

[6] For more on refining, see CRS Report R41478, *The U.S. Oil Refining Industry: Background in Changing Markets and Fuel Policies*, by Anthony Andrews, Robert Pirog, and Molly F. Sherlock.

[7] Mexican output fell sharply due to depletion at the giant Cantarell field. Efforts to offset the decline from other sources have been hampered by difficulties at the Mexican national oil company and restrictions on foreign investment.

[8] OPEC cut supply after oil prices started falling in the second half of 2008.

[9] For more on recent issues around infrastructure for oil imports from Canada, see CRS Report R41668, *Keystone XL Pipeline Project: Key Issues*, by Paul W. Parfomak et al.

[10] Note that because many European refineries were undergoing annual maintenance at the time, it delayed the full extent of the immediate physical impact of Libya shut-in exports. But they can still bid up crude prices as they look to secure supply now for use later in the year.

[11] It takes months or years to develop an oil field. Short term options are mostly limited to OPEC, which holds spare oil production capacity, or inventories held by companies or governments, which is discussed in greater detail below.

[12] CRS Report RS22204, *U.S. Trade Deficit and the Impact of Changing Oil Prices*, by James K. Jackson

[13] U.S. refiners' acquisition cost (RAC) for imported oil was $75.88 per barrel in 2010 versus $48.86 per barrel in 2005. The acquisition cost for domestic crude grew by nearly as much. See http://www.eia.doe.gov/dnav/pet/pet_pri_rac2_dcu_nus_a.htm.

[14] Nigel Gault, "Oil Prices and the U.S. Economy: Some Rules of Thumb," *IHS Global Insight Inc.*, February 24, 2011. Estimates from other forecasters of the Gross Domestic Product (GDP) impact of a $10 per barrel increase are similar, including those from the Federal Reserve according to reports (Robin Harding, "Oil Surge Puts Fragile US Recovery at Risk," *Financial Times*, February 24, 2011).

[15] James Hamilton, *Causes and Consequences of the Oil Shock of 2007-08*, Brookings Institute, Brookings Papers on Economic Activity, March 23, 2009, p. 40, *http://www* .brookings.edu/economics/bpea/~/media/Files/Programs/ES/BPEA/2009_spring_bpea_papers/2009_spring_ bpea_hamilton.pdf.

[16] Richard Newell, *What's the Big Deal About Oil? How We Can Get Oil Policy Right.*, Resources for the Future, 2006, p. 8, *http://www.rff.org/rff/News/Features* /upload/25007_1.pdf.

[17] Forecasters from a range of institutions have come to this conclusion, including EIA (*2011 Annual Energy Outlook*, Table 11, http://www.eia.gov/forecasts/aeo/tables_ref.cfm), IEA (*World Energy Outlook 2010* p. 134), OPEC (World Oil Outlook, Chapter 8, http://www.opec.org/opec_web/en/publications/340.htm), and BP (Energy Outlook 2030, p. 72, www.bp.com/energyoutlook2030).

[18] A dramatic increase in oil prices during the 1970s, substitution of oil with other fuels in non-transport sectors, conservation efforts, and greater domestic supply.

[19] CRS has produced a number of reports on U.S. interests in major oil producing countries which can be found at CRS.gov under the search terms "oil and foreign policy."

[20] The President's Materials Policy Commission, *Resources For Freedom*, July 2, 1952, p. 108.

[21] Mandatory Oil Import Program (MOIP), started in 1959, put a cap on how much foreign oil a refiner could import. Refiners were issued import allowance tickets that they could trade depending on their purchases of foreign oil. The program was managed to try to sustain domestic oil prices above $3/bbl.

[22] CRS Report RL30085, *Depressed Crude Oil Prices: Some Policy Options for Domestic Producers*, by Robert Bamberger, et al.

[23] Except for petroleum-based lubricants, which have a tariff of 84¢/bbl.

[24] NAFTA is the free trade agreement between the United States, Mexico, and Canada. GSP is a U.S. program designed to promote economic growth in developing countries and includes about 131 beneficiary nations and territories. Originally created by the Trade Act of 1974, the GSP's authorization expired at the end of 2010. S. 308 has been introduced in the 112th Congress to reauthorize the GSP until June 12, 2012. It would apply retroactively to imports since December 31, 2011. Congress regularly reauthorizes GSP for short periods. The program had expired seven times between 1993 and 2002; each time it was reauthorized and applied retroactively for the lapsed period. See http://www.ustr.gov/webfm_send/2465. AGOA was enacted in 2000 to promote stable sustainable economic growth and development in Sub-Saharan Africa. It has a broader list of products than GSP that eligible countries may export to the United States that are subject to zero import duty. Oil accounts for more than 90% of AGOA imports—mostly from Nigeria and Angola, both OPEC members (http://www.agoa.gov/resources/US_African_Trade_Profile_2009.pdf).

[25] Data provided by the International Trade Administration. Forgone revenue estimate assumes a scenario where the same number of barrels would have been imported from countries with waivers if such waivers did not exist. This is an assumption. Although the tariff amount is small, it is possible that its absence could have shifted oil trade flows (e.g., that less oil would have been imported to the United States from these particular countries if waivers did not exist).

[26] U.S. President (Clinton), "The Uruguay Round Trade Agreements, Texts of Agreements Implementing Bill, Statement of Administrative Action and Required Supporting Statements," 103rd Cong., 2nd sess., September 27, 1994, 103-316, Vol 2 (Washington: GPO, 1994), pp. 2418-2422.

[27] Ibid, p. 2418.

[28] Information in this paragraph and more background on oil exports from the United States can be found in CRS Report R40120, *U.S. Oil Exports*, by Robert Bamberger.

[29] The Administration target is to cut oil imports by one-third in about a decade, relative to their levels when President Obama took office. The U.S. imported around 11 Mb/d net in 2008, so a one-third reduction would be 3.6 Mb/d. The target and plan were presented in President Obama's March 30, 2011 speech on energy, available at http://www.whitehouse.gov/the-press-office/2011/03/30/remarks-president-americas-energy-security.

[30] CRS Report R40110, *Biofuels Incentives: A Summary of Federal Programs*, by Brent D. Yacobucci.

[31] Congressional Budget Office, *Using Biofuel Tax Credits to Achieve Energy and Environmental Policy Goals*, July 2010, http://www.cbo.gov/ftpdocs/114xx/doc11477/07-14-Biofuels.pdf.

[32] CRS Report R40445, *Intermediate-Level Blends of Ethanol in Gasoline, and the Ethanol "Blend Wall"*, by Brent D. Yacobucci.

[33] For more information on the Deepwater Horizon Oil Spill and issues for Congress, see CRS Report R41407, *Deepwater Horizon Oil Spill: Highlighted Actions and Issues*, by Curry L. Hagerty and Jonathan L. Ramseur.

[34] According to the Council of Economic Advisors, CARS may have raised third quarter 2009 gross domestic product (GDP) growth by around 0.2 percentage points (annualized rate). Council of Economic Advisers, "Economic Analysis of the Car Allowance Rebate System," September 10, 2009, p. 12. Note, the United States was in recession from December 2007 until June 2009.

[35] Ibid, p. 13.

[36] According to National Highway Safety Administration (NHTSA), CARS cost approximately $3.0 billion, but provided future fuel savings for consumers who participated of only an expected $1.3 to $1.9 billion (based on applying a 7% and a 3% discount rate respectively to calculate present value of future savings). NHTSA, *Consumer Assistance to Recycle and Save Act of 2009: Report to the House Committee on Energy and Commerce, the Senate Committee on Commerce, Science, and Transportation and the House and Senate Committees on* Appropriations, December 2009, p. 46, *http://www.cars.gov/files/official-information/CARS-Report-to-Congress.*

This may still be the case even if one assumes a cost of the greenhouse gas emissions avoided: assuming $20 per ton, CARS saved another roughly $0.2 billion through avoided emissions (p. 49). The value of reducing other air pollutants was estimated at $0.2 billion (p. 53). Critics pointed out that the program could have been more cost effective had it required a wider difference in the fuel efficiency between cars traded in and new cars purchased.

[37] CRS Report R40654, *Accelerated Vehicle Retirement for Fuel Economy: "Cash for Clunkers"*, by Brent D. Yacobucci and Bill Canis.

[38] Edmunds.com, "Cash for Clunkers Results Finally In: Taxpayers Paid $24,000 per vehicle sold, Report Edmunds.com," October 28, 2009.

[39] The amount of oil held in the SPR is frequently described as a fraction of U.S. daily average imports. Were SPR oil released, it can only be pumped at a maximum rate of 4.4 Mb/d for up to 90 days. Then the drawdown rate begins to decline as storage caverns are emptied. At full capacity, the SPR can send oil to the market for about 180 days.

[40] The Heating Oil Reserve is held to supplement commercial heating oil supplies should the heavily heating oil-dependent North East region be hit by a severe heating oil supply disruption. Since its creation in 2000, the Heating Oil Reserve has not been used for emergency purposes. Oil has been sold from the Heating Oil Reserve for budgetary reasons and to convert inventories to fuel with lower sulfur content. For more details, see http://www.fossil.energy.gov/programs/reserves/heatingoil/index.html.

[41] The United States and other members of the Organisation for Economic Co-operation and Development (OECD) agreed to hold strategic reserves and coordinate their use through the 1974 Agreement On An International Energy Program treaty (*http://www* .iea.org/about/docs/IEP.PDF). The treaty established the International Energy Agency (IEA) to carry out this coordination. IEA members are required to hold 90 days worth of oil imports in reserve and coordinate their use. The United States counts commercial inventories to satisfy the remainder of the 90-day requirement not met by the SPR.

[42] International Energy Agency, *IEA Response System for Oil Supply Emergencies*, 2011, http://www.iea.org/textbase/nppdf/free/2011/response_system.pdf.

[43] International Energy Agency, *Fact Sheet: IEA Stocks and Drawdown Capacity*, February 25, 2011, *http://www.iea.org/files/Potential_IEA_Stockdraw_Capacity.pdf.*

In: U.S. Oil Sources
Editor: Natalie B. Kelleher
ISBN: 978-1-62100-415-8

Chapter 2

THE U.S.-CANADA ENERGY RELATIONSHIP: JOINED AT THE WELL

Paul W. Parfomak and Michael Ratner

SUMMARY

The United States and Canada, while independent countries, effectively comprise a single integrated market for petroleum and natural gas. Canada is the single largest foreign supplier of petroleum products and natural gas to the United States—and the United States is the dominant consumer of Canada's energy exports. The value of the petroleum and natural gas trade between the two countries totaled nearly $100 billion in 2010, helping to promote general economic growth and directly support thousands of energy industry and related jobs on both sides of the border. Increased energy trade between the United States and Canada—a stable, friendly neighbor—is viewed by many as a major contributor to U.S. energy security. The U.S.-Canada energy relationship is increasingly complex, however, and is undergoing fundamental change, particularly in the petroleum and natural gas sectors.

Congress has been facing important policy questions in the U.S.-Canada energy context on several fronts, including the siting of major cross-border pipelines, increasing petroleum supplies from Canadian oil sands, increasing natural gas production from North American shales, and the construction of new facilities for liquefied natural gas (LNG) exports. Legislative proposals in the 112th Congress could directly influence these developments. These proposals include H.R. 1938, which would expedite consideration of the Keystone XL pipeline proposal, H.R. 909, which would encourage petroleum and natural gas production on the outer continental shelf and in the Arctic National Wildlife Refuge, and S. 304, which would support a program to train workers involved with oil and gas infrastructure in Alaska. Other proposals in Congress affecting hydraulic fracturing operations for natural gas production, offshore drilling, or U.S. oil shale development could also affect the U.S.-Canada energy relationship.

Traditionally, the energy trade between the United States and Canada, while intertwined, has been uncomplicated—taking the form of a steadily growing southward flow of crude oil and natural gas to markets in the U.S. Midwest and Northeast. But recent developments have greatly complicated that energy relationship creating new competition and interconnections. Consequently, while energy policies in one country have always inevitably affected the other, their cross-cutting effects in the future may not be widely understood and, in some cases, may be largely unanticipated. For example, policies affecting U.S. shale gas production could affect North American natural gas prices overall, and thus, the costs of producing petroleum from oil sands (which requires large volumes of natural gas for heating). Changing oil sands costs could, in turn, affect Canadian petroleum supplies to the United States, affecting north-south pipeline use and changing U.S. petroleum import requirements from overseas. Changing natural gas prices would also change the economics of Arctic natural gas, however, and influence the development of the Arctic natural gas pipelines, which could provide an alternative source of economic natural gas for oil sands production in Alberta. How such scenarios could play out in reality is open to debate, but they illustrate the tangled web policymakers in both countries must navigate as they consider future energy, environmental, and transportation decisions.

As Congress debates legislative proposals affecting the petroleum and natural gas industries, it may be helpful to consider these proposals in the broadest possible North American context, recognizing that the energy sector in Canada may be moved in one direction or another based on policies in Washington, DC. To date, the judgment of Congress has favored a growing U.S.-Canada energy partnership—but ensuring that this relationship continues to be as mutually beneficial as possible will likely remain a key oversight challenge for the next decades.

INTRODUCTION

Canada has long been the United States' most important energy partner. Canada is the single largest foreign supplier of petroleum products, natural gas, and electric power to the United States—and the United States is the dominant consumer of Canada's energy exports. Canada is also the primary recipient of U.S. energy exports. The value of the energy trade between the two countries totaled nearly $100 billion in 2010, helping to promote general economic growth and directly support thousands of energy industry and related jobs on both sides of the border. Increased energy trade between the United States and Canada—a stable, friendly neighbor—is viewed by many as a major contributor to U.S. energy security. The U.S.-Canada energy relationship is increasingly complex, however, and is undergoing fundamental change, particularly in the petroleum and natural gas sectors.

Congress has been facing important policy questions in the U.S.-Canada energy context on several fronts, including the siting of major cross-border pipelines, increasing petroleum supplies from Canadian oil sands, increasing natural gas production from North American shales, and the construction of new facilities for liquefied natural gas (LNG) exports. Legislative proposals in the 112th Congress could directly influence these developments. For example, H.R. 1938 would direct the President to expedite the consideration and approval of the Keystone XL pipeline linking Canadian oil sands production to refineries in the Gulf of Mexico. H.R. 909 would encourage petroleum and natural gas production on the outer continental shelf, would prescribe requirements for coordination with adjacent states regarding associated pipeline construction, and would allow production of petroleum and natural gas from the Arctic National Wildlife Refuge, among other provisions. S. 304 would support a program to train workers in the construction, operation, maintenance, and performance of all related environmental processes involving oil and gas infrastructure in Alaska. Other proposals in Congress affecting hydraulic fracturing operations for natural gas production, offshore drilling, or U.S. oil shale development could also affect the U.S.-Canada energy trade.

While specific energy policy issues arising in the United States and Canada may appear to be independent of one another, many have important physical, economic, and environmental links. Thus, U.S.-Canada energy policies established in one context may have important implications in others.

This report provides an overview of the U.S.-Canada energy trade, with a focus on petroleum and natural gas.[1] It summarizes important trends in both of these sectors and identifies key connections among these trends. Finally, the report discusses possible implications for the U.S.-Canada energy relationship going forward, highlighting considerations for Congress as it continues its oversight of the energy industry and considers new energy legislation.[2] Although the report raises environmental issues in specific contexts, a broad discussion of environmental impacts from North American energy production and consumption is beyond its scope.

U.S.-CANADA ENERGY TRADE AND INFRASTRUCTURE[3]

In 2010, 30% of primary energy consumed in the United States was imported (**Figure 1**).[4] Approximately 29% of these energy imports were Canadian petroleum (18%) and Canadian natural gas (11%).[5] Taken together, Canadian petroleum and natural gas accounted for 9% of total U.S. primary energy consumption in 2010, the largest contribution from any one foreign supplier.

The value of energy imports from Canada in 2010 was $83.6 billion, accounting for about 3.5% of all U.S. imports of goods and services that year.[6] These payments were primarily for petroleum and natural gas (**Figure 2**). In addition to these imports, the United States exported $13.5 billion worth of energy commodities to Canada, including $5.3 billion of refined petroleum products. The North America Free Trade Agreement's (NAFTA) extensive energy provisions have facilitated energy trade between the United States and Canada, underscoring the importance of this trade for both countries. NAFTA states, in part, "it is desirable to strengthen the important role that trade in energy and basic petrochemical goods plays in the free trade area...."[7]

By virtue of NAFTA and the private sector orientation of the energy sectors in both countries, U.S. and Canadian companies have become integrated in the development, production, transportation, and marketing of petroleum and natural gas. Joint ventures between U.S. and Canadian companies on petroleum and natural gas projects are common. These close connections, and geographic proximity, have led the U.S. and Canadian energy markets to be viewed as one.

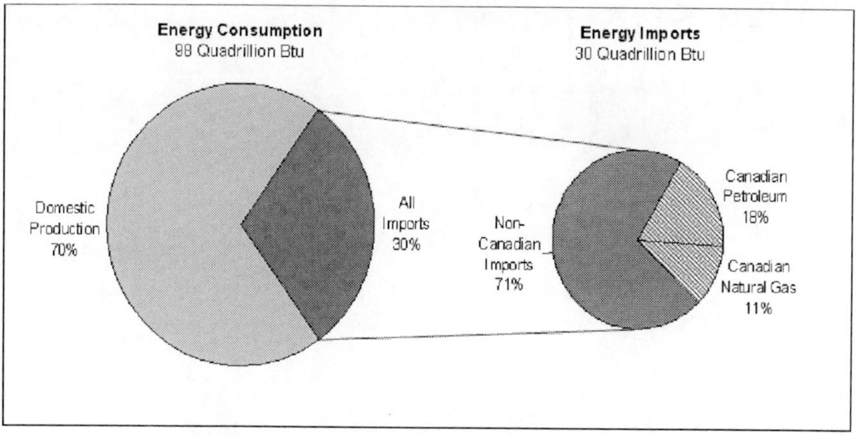

Source: U.S. Energy Information Administration, *Monthly Energy Review*, April 2011, pp. 3, 41, 45, and 70.

Notes: Btu = British thermal unit. One quadrillion Btu equals about 172 million barrels of oil or 974 billion cubic feet of natural gas.

Figure 1. 2010 U.S. Primary Energy Consumption and Sources.

Total = $83.6 billion

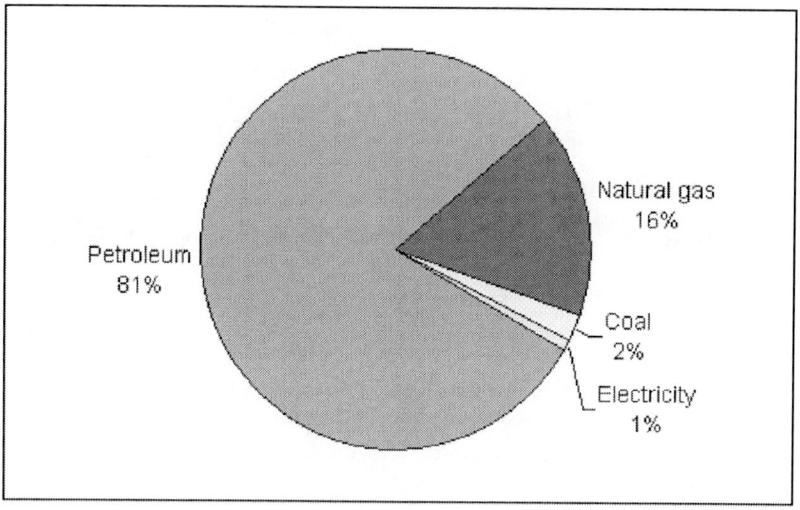

Source: U.S. Census Bureau, "U.S. Imports from Canada by 5-digit End-Use Code 2002 – 2010," March 15, 2011, http://www.census.gov/foreign-trade/statistics/product/enduse/imports/c1220.html.

Figure 2. Value of U.S. Energy Imports from Canada 2010.

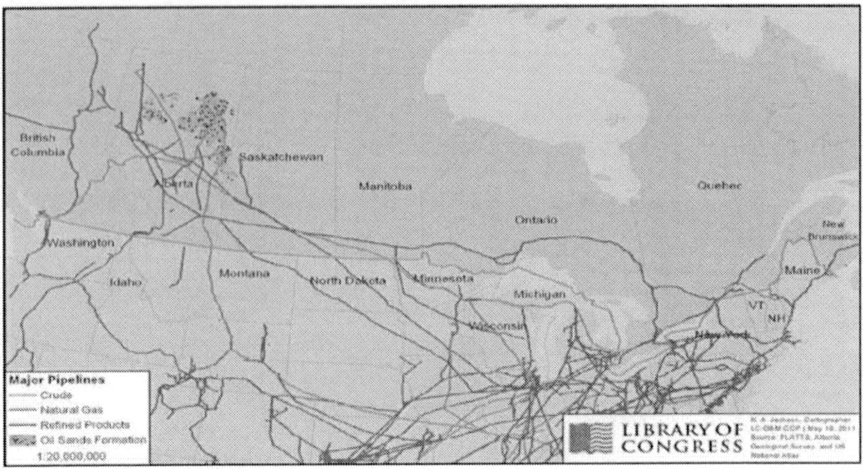

Source: Compiled by the Library of Congress's cartography division.

Figure 3. Major U.S.-Canada Energy Commodity Pipelines.

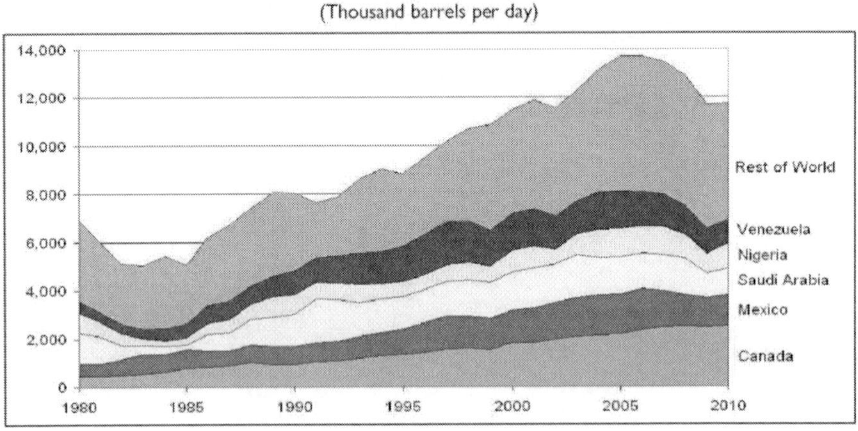

Source: U.S. Energy Information Administration, Petroleum & Other Liquids, U.S. Imports by Country of Origin," Online tables, July 29, 2010, http://www .eia.gov/dnav/pet/ pet_move_impcus_a2_nus_ep00_im0_mbblpd_a.htm.

Notes: In 2010, total U.S. gross imports were about 11.8 Mbpd, down from a peak of 13.7 Mbpd in 2005. During that period, U.S. oil exports increased from 1.2 Mbpd to 2.3 Mbpd, nearly all in the form of refined petroleum products.

Figure 4. Gross U.S. Oil Imports 1980-2010.

Energy Pipeline Infrastructure

The United States and Canada are connected by high capacity pipelines carrying crude oil, other petroleum products, and natural gas. As **Figure 3** shows, pipelines originating in Canadian supply basins are linked to major markets across the United States, comprising a large part of the North American pipeline system. There are five major Canadian petroleum export pipelines, discussed later in this report. Canadian natural gas exports cross into the United States at 25 entry points across the length of the border through major and minor pipelines. The primary receiving states are Idaho (21%), New York (20%), Montana (15%), North Dakota (15%), and Minnesota (15%).[8] While pipelines carry most Canadian energy exports to the United States, significant volumes are also transported by truck, train, and marine vessel.[9]

KEY DEVELOPMENTS IN U.S.-CANADA ENERGY

Although the United States and Canada have a long-established trade relationship in petroleum and natural gas—both have been shipped across the border since the late 1800s—several aspects of that relationship have been undergoing a transformation in recent years. These changes include the rapid growth in petroleum supplies from Canadian oil sands, the siting of major cross-border petroleum pipelines, renewed attempts to commercialize Arctic natural gas, a sharp rise in natural gas production from U.S. shales, and the development of new liquefied natural gas (LNG) facilities. In each of these areas new technology and infrastructure investments may have a significant effect on the balance of energy supply, demand, and trade between the United States and Canada. In some cases, they may also create new competition between the two countries in developing specific mineral resources and infrastructure projects.

Growth in Petroleum Supplies from Canadian Oil Sands

In 2010, Canada was the largest supplier of imported petroleum to the United States.[10] Of the 11.8 million barrels per day (Mbpd) the United States imported last year, Canada supplied 2.5 Mbpd (22%), more than the imports from the next two largest suppliers combined—Mexico and Saudi Arabia (**Figure 4**). Canadian imports have grown fairly steadily since the early 1980s and are expected to continue growing as new U.S.-Canada pipeline capacity is added and Canadian petroleum resource development expands. The Canadian Association of Petroleum Producers (CAPP) projects crude oil output to increase more than 50% from 2010 to 2025, with most of this production destined for the United States.[11] As noted earlier, the United States also exports a limited amount of petroleum to Canada, mostly as refined products.

Petroleum production from oil sands is the key driver behind the growth in Canadian petroleum exports. Oil sands are mixtures of sand, water, and bitumen.[12] When oil sands, also known as tar sands, are included, Canada's petroleum reserves rank second in the world behind Saudi Arabia's.[13] Canada has an estimated 143.1 billion barrels of petroleum reserves, of which 81% are from oil sands. According to CAPP projections, by 2025, oil sands will account for about 80% of total Canadian oil production, up from 50% currently.[14]

Notwithstanding its rapid growth, petroleum production from Canadian oil sands is controversial because it has significant environmental impacts, including emissions of greenhouse gases during extraction and processing (that exceed emissions from conventional oil production), disturbance of mined land, and impacts on wildlife and water quality.[15] Since bitumen in oil sands cannot be pumped from a conventional well, it must be mined, usually using strip mining or open pit techniques, or the oil can be extracted using underground heating methods, which require large amounts of water and natural gas (for heating).[16] The magnitude of the environmental impacts of oil sands production, in absolute terms and compared to conventional oil production, has been the subject of numerous, and sometimes conflicting, studies and policy papers.[17]

Table 1. Major U.S.-Canadian Petroleum Import Pipelines

Pipeline	Operator	Status	Capacity (bpd)
Enbridge Mainline	Enbridge	Operating	1,868,000
Express	Kinder Morgan	Operating	280,000
Trans Mountain	Kinder Morgan	Operating	300,000
Alberta Clipper	Enbridge	Began operating in 2010	450,000
Keystone	TransCanada	Began operating in 2010	435,000
Keystone XL	TransCanada	Proposed	830,000
TOTAL			**4,163,000**

Source: CAPP, June 2010, p. 19; U.S. Department of State, April 15, 2011, p. 1-4.

Development of New Cross-Border Petroleum Pipelines

Cross-border pipeline infrastructure for Canadian petroleum exports to the United States has been growing rapidly. Five major pipelines with a combined capacity of 3.3 Mbpd currently link Canadian petroleum producing regions to markets in the United States (**Table 1**). Two of these pipelines, Alberta Clipper and Keystone, with a combined capacity of just under 0.9 Mbpd (26% of the total) began service last year. A permit application for a sixth pipeline, Keystone XL, which would add an additional 0.8 Mbpd of capacity, is in the final stages of review by the U.S. State Department.[18] If approved and constructed, Keystone XL would bring Canada's total U.S. petroleum export capacity to over 4.1Mbpd, enough capacity to carry over 34% of U.S. petroleum imports in 2010.[19] Given that Canada actually supplied the United States 2.5 Mbpd in 2010, large increases in Canadian supply will ultimately be possible, although the industry anticipates significant excess pipeline capacity for the next decade.[20] In addition, several large pipeline projects are proposed within the United States to increase movements of Canadian petroleum to key U.S. market hubs, including refineries in the Midwest and on the Gulf Coast that employ complex technology in order to process "heavy" crude oils like those from Canada, Mexico, and Venezuela. Note that whether a pipeline is located in one country or the other has little bearing on its ownership; Kinder Morgan is a U.S. company while Enbridge and TransCanada are Canadian companies (**Table 1**).

The recent expansion of petroleum pipelines from Canada has generated considerable controversy in the United States. Proponents of these pipelines, including Canadian government agencies, petroleum industry stakeholders, and pipeline construction workers, have based their public interest justifications primarily on increasing the diversity of the U.S. petroleum supply and on expected economic benefits to the United States, including near-term job creation associated with pipeline construction and operation. Opponents, primarily environmental groups and affected communities along pipeline routes, have objected to these projects principally on the grounds that Canadian oil sands development has negative environmental impacts and that it promotes continued U.S. dependency on fossil fuels. Some opponents also argue that, given the excess capacity anticipated in the existing Canadian petroleum pipelines noted above, additional pipelines are not needed. These issues have come into particular focus in the context of the Keystone XL pipeline proposal, which applied for a Presidential Permit from the U.S.

Department of State in 2008.[21] The State Department expects to make a decision regarding this permit by the end of 2011. [22] H.R. 1938 (Sec. 3) would require this decision by November 1, 2011.

Arctic Natural Gas Pipeline Proposals

The Arctic region has substantial natural gas resources. For example, the U.S. Geological Survey (USGS) estimates that conventional natural gas reserves just on Alaska's North Slope potentially exceed 100 trillion cubic feet (Tcf), over four times the total annual gas consumption of the United States.[23] The Mackenzie River Delta region in the Canadian Arctic contains an estimated 40 Tcf of natural gas.[24] The USGS's assessment of undiscovered conventional natural gas resources across the entire Arctic region concluded that over 1,600 Tcf of additional natural gas resources remain to be found, much of it under Canadian and U.S. territory.[25] Both the United States and Canada have long recognized the potential of these natural gas resources and have pursued policies to develop them. Principal among these policies has been promoting the construction of natural gas pipelines from the Arctic to markets in the lower-48 United States.

While Arctic natural gas pipeline projects have been on and off the drawing board for decades, serious interest in Arctic natural gas pipeline revived around 2000 because of accelerated growth in U.S. natural gas demand, rising natural gas prices, and increased importation of liquefied natural gas (LNG) from overseas. Moreover, many industry analysts expected a U.S. policy of carbon dioxide control could further increase natural gas demand for electric power generation and, possibly, transportation fuel. These factors led both U.S. and Canadian officials to restart the process of Arctic natural gas pipeline development.[26]

Important milestones in Arctic pipeline activity were Alaska's 2008 award to TransCanada of a license to build an Alaska natural gas transportation system from Prudhoe Bay, AK, through Canada to the lower-48 states, the concurrent announcement of a competing pipeline proposal (Denali) along a similar route, and revival of a third proposal for an all-Canada pipeline originating in the Mackenzie Delta (**Figure 5**). Since large sections of either of the proposed Alaska natural gas pipelines would pass through Canadian territory, Canada has cooperated with the United States on their development. However, because the Mackenzie pipeline would commercialize a major new source of North American natural gas, and would draw on a limited pool of construction resources and materials available for such a project, it has been viewed by some as a direct competitor to an Alaska gas pipeline.[27]

Notwithstanding recent development progress, there have been many obstacles to Arctic natural gas pipelines—most notably natural gas prices in the lower-48 states, the primary market for Arctic natural gas. As discussed below, a rapid and largely unanticipated increase in natural gas production from U.S. shales has lowered natural gas price forecasts for the foreseeable future. Given this drop in prices, Arctic natural gas projects may not be economically viable at present. In March 2011, Canadian authorities provisionally approved the Mackenzie pipeline project, although some analysts believe it may not be constructed without new government subsidies.[28] In May 2011 the developers of the Denali pipeline proposal discontinued the project, citing a lack of commitment to contract for pipeline capacity among potential Arctic gas producers (two of which—BP and ConocoPhillips—were Denali sponsors).[29] TransCanada officials have stated that they remain committed to developing their Alaska pipeline project, although some industry analysts are skeptical.[30] If either the TransCanada or Mackenzie pipeline is ultimately constructed and begins transporting natural gas to lower-48 markets, it could have a significant impact on U.S. energy prices, energy security, and emissions of carbon dioxide.

Source: Adapted from CRS Report RL31278, *Arctic National Wildlife Refuge: Background and Issues*, Figure 6, based on Energy Department maps.

Figure 5. Arctic Natural Gas Pipelines Proposals.

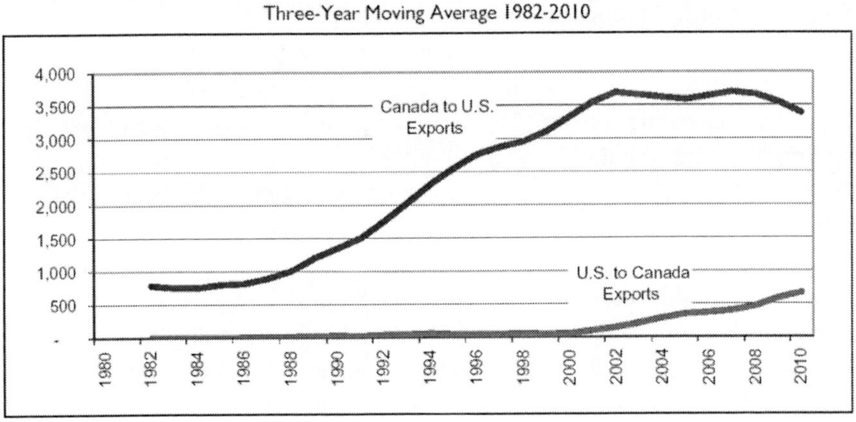

Source: U.S. Energy Information Administration, "Annual U.S. Natural Gas Pipeline Imports From Canada," and "Annual U.S. Natural Gas Pipeline Exports to Canada," Internet tables, April 29, 2011, http://www.eia.gov/dnav/ng/hist/n9102cn2a.htm and http://www.eia.doe.gov/dnav/ng/hist/n9132cn2A.htm.

Figure 6. Canada-U.S. Natural Gas Exports (bcf).

Declining U.S. Natural Gas Imports[31]

Canada is the dominant foreign supplier of natural gas to the United States. Of the 3,683 billion cubic feet (bcf) of natural gas the United States imported in 2010, Canada supplied 3,222 bcf (88%), almost 20 times the next largest supplier. This level of Canadian natural gas exports to the United States comprised approximately 13% of total U.S. natural gas consumption last year. Canadian natural gas exports to the United States saw rapid growth beginning in the mid-1980s, rising more than 400% between 1984 and 2002. These imports have begun to decline, however, due to increases in U.S. imports of LNG from overseas and, more recently and importantly, due to increases in domestic natural gas production. At the same time, the United States has begun to export significant volumes of U.S.-produced natural gas to markets in eastern Canada, displacing Canadian supplies (**Figure 6**). In 2010, the United States exported 738 billion cubic feet of natural gas to Canada, mostly via Michigan. In May, the U.S. Federal Energy Regulatory Commission approved a new pipeline from Marcellus shale gas fields in the United States to Canada, potentially increasing northward natural gas exports.[32]

The rise in U.S. domestic natural gas supplies has been driven by an unanticipated growth in natural gas production from shale, a widespread type of geologic formation that often holds large quantities of natural gas but poses technical challenges for extraction. In recent years, energy companies have overcome these challenges, making large natural gas resources in U.S. shales commercially available. For example, in its 2011 *Annual Energy Outlook*, the Energy Information Administration more than doubled its estimate of shale gas resources to 827 Tcf from its 2010 *Annual Energy Outlook* estimate of 347 Tcf—due primarily to a re-evaluation of shale gas supplies.[33] Because U.S. shale gas reserves are located close to major natural gas markets, U.S. shale gas has an advantage over traditional supply basins in Western Canada and the U.S. Gulf Coast. Large shale gas reserves are also found in Canada, although they are several years behind the development of U.S. reserves because of limited pipeline infrastructure and because transportation costs make them less competitive than other North American supplies.[34]

The reversal of Canada-U.S. natural gas export trends is driving fundamental changes in the North American natural gas industry. For example, due to sharply declining long-haul contract volumes on its Canadian natural gas pipelines to the United States, TransCanada has had to drastically restructure its pipeline tariffs to maintain the economic viability of certain lines.[35] Likewise, as discussed earlier, shale gas plays have hurt prospects for Arctic natural gas development in both Alaska and Arctic Canada. On the other hand, low natural gas prices benefit the production of petroleum from oil sands, which requires large volumes of natural gas for extraction and processing. It remains to be seen how the development of additional North American natural gas reserves and associated pipeline infrastructure will turn out over time as Canada seeks to develop its own resources (with the help of U.S. companies) in response to U.S. shale gas developments. At least in the near-term, however, it appears that recent trends of reduced U.S. imports of Canadian natural gas will continue.

Liquefied Natural Gas Import-Export Terminals

Liquefied natural gas (LNG) shipments from overseas historically have played a minor role in North American energy markets. However, in reaction to rising natural gas prices in the early 2000s and a fear of impending shortages of pipeline natural gas, demand for LNG imports to the United States was expected to increase.[36] To meet this anticipated growth in LNG demand, developers expanded existing LNG terminals and constructed or proposed numerous new LNG import terminals in both the United States and

Canada. In the United States, between 2001 and 2011, three existing LNG import terminals were expanded, seven new LNG terminals were constructed, and sixteen were approved but not constructed. In Canada during this period, one new LNG import terminal was constructed and two more were approved. Because Canada's new LNG terminals would serve the same Northeast markets as several of the proposed U.S. terminals, there was considerable competition among the developers of these projects. In one instance, this competition even reportedly created diplomatic tension between the two countries.[37] But the building boom in LNG terminals was premature. As North American natural gas supplies from shale plays rapidly increased, the anticipated rise in demand for LNG imports did not materialize, leaving most of the new LNG import capacity unutilized.

In a somewhat ironic turn of events given the origins of the aforementioned LNG terminal building boom, some terminal owners and developers are now proposing to *export* North American natural gas to China, Japan, and other foreign buyers. At least two groups have proposed Canadian LNG export terminals in British Columbia, anticipating significant natural gas supplies from shales in Western Canada. Many analysts view such exports as the only way to economically develop gas reserves from these western shales, which might otherwise not be competitive with U.S. natural gas supplies. Likewise at least three U.S. developers have filed applications for LNG export facilities (modifying existing import terminals), and more such filings are anticipated. On May 20, 2011, the U.S. Department of Energy issued its first conditional authorization for LNG exports from the Sabine Pass LNG Terminal in Louisiana to non-NAFTA countries.[38] If U.S. and Canadian LNG export projects are developed, LNG developers in the two countries could again find themselves in competition, only this time in seeking to supply overseas LNG buyers. The potential effects on North American natural gas prices are difficult to predict, but exports to foreign markets would create upward price pressures.[39] Furthermore, a large LNG export trade might limit U.S. natural gas exports to Canada in the future.

CONSIDERATIONS FOR CONGRESS

The United States and Canada, while independent countries, effectively comprise a single integrated market for petroleum and natural gas. These markets are physically linked by billions of dollars of transportation and refining infrastructure, and are economically linked by direct participation in the same regional and global energy markets. Canada is the largest foreign supplier of energy to the United States and will continue to be for the foreseeable future. The United States depends on Canada for oil and natural gas supplies that it cannot currently produce itself. As the primary supplier of U.S. imports of petroleum and natural gas, Canada is viewed as a stabilizing factor for U.S. energy supplies; although petroleum prices are set in a global market, the likelihood that Canada would cut off oil and natural gas supplies is remote. But Canada is equally dependent upon the United States to buy energy exports that might not easily find a market elsewhere due to geographical constraints. The United States is also a critical supplier to Canada of refined petroleum products. U.S.-based companies invest heavily in assets and energy resources in Canada and vice versa. Although individual companies in both countries may compete for specific energy opportunities (e.g., LNG terminals), the overall energy relationship between the United State and Canada is mutually beneficial.

Traditionally, the energy trade between the United States and Canada, while intertwined, has been uncomplicated—taking the form of a steadily growing southward flow of crude oil and natural gas to markets in the U.S. Midwest and Northeast. But recent increases in oil sands and shale gas production, expansion of cross-border pipeline capacity, prospects for LNG exports, and renewed interest in Arctic natural gas have greatly complicated that energy relationship creating new competition and interconnections. Consequently, while energy policies in one country have always inevitably effected the other, their cross-cutting effects in the future may not be widely understood and, in some cases, may be largely unanticipated. For example, policies affecting U.S. shale gas production could affect North American natural gas prices overall, and thus, the costs of producing petroleum from oil sands (which requires large volumes of natural gas for heating). Changing oil sands costs could, in turn, affect Canadian petroleum supplies to the United States, affecting north-south pipeline use and changing U.S. petroleum import requirements from overseas. Changing natural gas prices would also change the economics of Arctic natural gas, however, and influence the development of the Arctic natural gas pipelines, which could provide an alternative source of economic natural gas for oil sands production in Alberta. How such scenarios could play out in reality is open to debate, but they illustrate the tangled web policymakers in both countries must navigate as they consider future energy, environmental, and transportation decisions.

As Congress debates legislative proposals affecting the petroleum and natural gas industries, it may be helpful to consider these proposals in the broadest possible North American context, recognizing that the energy sector in Canada may be moved in one direction or another based on policies in Washington, DC. For example, developers are already pursuing western Canadian routes for petroleum exports to Asia as an alternative to U.S. exports, especially if the latter should fail to grow as expected.[40] Ultimately, the energy market effects of specific energy policies and projects must be weighed against their broader economic value, energy security implications, and environmental impacts. To date, the judgment of Congress has favored a growing U.S.-Canada energy partnership—but ensuring that this relationship continues to be as mutually beneficial as possible will likely remain a key oversight challenge for the next decades. If the balance tips the other way—either in the eyes of developers or the federal government—Congress may need to reconsider its position on key energy and related initiatives to meet the United States' long-term policy objectives.

ACKNOWLEDGMENTS

Neelesh Nerurkar, Specialist in Energy Policy, contributed to this report.

End Notes

[1] Although regionally significant, Canadian electricity exports to the United States are small on a national basis. In 2009, Canadian imports comprised approximately 4,000 terrawatt-hours, or about 1% of U.S. electricity consumption. Canada also supplies approximately one-third of the uranium used in U.S. nuclear power plants. Issues related to the U.S.-Canada electric grid and nuclear power sectors are important, but beyond the scope of this report.

[2] For broader analyses of the U.S. relationship with Canada, see CRS Report 96-397, *Canada-U.S. Relations*, coordinated by Carl Ek and Ian F. Fergusson, or CRS Report RL33087, *United States-Canada Trade and Economic Relationship: Prospects and Challenges*, by Ian F. Fergusson.

[3] The data used in this report are for gross imports not net imports (imports less exports) unless otherwise specified.

[4] "Primary energy" includes all forms of energy when they are initially consumed. For example, natural gas can be converted to a liquid and consumed in a car. Natural gas, in this case, is the primary energy, while the liquid fuel is the secondary energy. Renewable forms of electricity generation, such as wind or solar, are considered primary energy, while electricity produced by natural gas or coal is not. Nuclear fuel is not included as energy in this report. Imports of electricity from nuclear and other fuel sources is included, but not broken out.

[5] "Petroleum" includes crude oil, fuel oil, natural gas liquids, and refined petroleum products (e.g., gasoline).

[6] U.S. Census Bureau, "U.S. Exports to Canada by 5-digit End-Use Code 2002-2010," March 15, 2011, http://www.census.gov/foreign-trade.

[7] North American Free Trade Agreement, Chapter Six: Energy and Basic Petrochemicals, Article 601.2, 1994.

[8] U.S. Energy Information Administration, "U.S. Natural Gas Imports by Point of Entry," Internet table, May 31, 2011, *http://www.eia.gov/dnav/ng/ng* _move_poel_a _EPG0_IRP_Mmcf_a.htm.

[9] Transport Canada, "Table RO18: International Trade Volume Shipped by Trucks, by Commodity Groups, 2008 and 2009," July 29, 2010, http://www.tc.gc.ca/eng/policy/report-aca-anre2009-2478.htm; "Table RA12: Value of Rail Exports and Imports by Commodity, 2000 – 2009," August 6, 2010, http://www.tc.gc.ca/eng/policy/report-aca-anre2009-2433.htm; "Table M29: Total Marine Imports/Exports by Commodities, 2008," July 29, 1010, http://www.tc.gc.ca/eng/policy/report-aca-anre2009-2520.htm.

[10] For additional analysis of U.S. petroleum imports, see CRS Report R41765, *U.S. Oil Imports: Context and Considerations*, by Neelesh Nerurkar.

[11] Canadian Association of Petroleum Producers (CAPP), "Crude Oil: Forecast, Markets & Pipelines," Calgary, AB, June 2010, p. 27, *http://www.capp.ca* /GetDoc.aspx ?DocId=173003.

[12] Bitumen is a black, sticky, and viscous substance that naturally occurs as a byproduct of decomposed organic materials. In the United States, bitumen is also known as asphalt, a binder mixed with aggregate in road surfaces.

[13] BP, *BP Statistical Review of World Energy*, June 2011, p. 6.

[14] CAPP, June 2010, p. 28.

[15] For more analysis of oil sands and their environmental impacts, see CRS Report RL34258, *North American Oil Sands: History of Development, Prospects for the Future*, by Marc Humphries.

[16] U.S. Bureau of Land Management, "About Tar Sands," web page, January 11, 2011, http://ostseis.anl.gov/guide/ tarsands/index.cfm; Cecilia Jamasmie, "The Challenges and Potential of Canada's Oil Sands," *Mining*, September-October 2010, pp. 7-8.

[17] For an example of contrasting views, see IHS CERA Inc., *Oil Sands, Greenhouse Gases, and US Oil Supply, Getting the Numbers Right*, 2010; and Natural Resources Defense Council, "Setting the Record Straight: Lifecycle Emissions of Tar Sands," November 2010.

[18] U.S. Department of State, Supplemental Draft Environmental Impact Statement for the Keystone XL Oil Pipeline Project, April 15, 2011. p. 1-4. An initial capacity of 700,000 bpd may be raised to 830,000 bpd by increasing the pumping capacity. The Keystone XL project had applied to the U.S. Pipeline Hazardous Materials Safety Administration to operate at slightly higher pressure than permitted in standard regulations, which would have enabled a 900,000 bpd capacity, but it withdrew its applications for such a Special Permit in August 2010.

[19] TransCanada has proposed a pipeline spur from the Keystone XL pipeline to the Bakken oil shale field in Montana, North Dakota, and South Dakota, known as the Bakken Marketlink Project. Although this project would be exclusively in the United States, it would be owned by TransCanada, again emphasizing the role companies play in the development each country's oil and natural gas projects.

[20] CAPP, June 2010, p.23.

[21] The construction, connection, operation, and maintenance of a pipeline connecting the United States with a foreign country requires executive permission through a Presidential Permit under Executive Orders 11423 and 13337.

[22] For additional analysis of the Keystone XL pipeline, see CRS Report R41668, *Keystone XL Pipeline Project: Key Issues*, by Paul W. Parfomak et al.

[23] David W. Houseknecht, U.S. Geological Survey, *Conventional Natural Gas Resource Potential, Alaska North Slope*, Open File Report 2004-1440, December 13, 2004; U.S.

Energy Information Administration, *Annual Energy Outlook 2009*, DOE/EIA-0383(2009), March 2009, p. 109.

[24] U.S. Geological Survey, "Assessment of Undiscovered Oil and Gas Resources of the Mackenzie Delta Province, North America, 2004," Fact Sheet 2006-3002, March, 2006.

[25] U.S. Geological Survey, *USGS Arctic Oil and Gas Report*, Fact Sheet, July 2008.

[26] U.S. Energy Information Administration, *Natural Gas Monthly*, August 2009, Table 2. Full initial pipeline capacity is 4.5 billion cubic feet per day.

[27] See, for example: James Irwin, "Alaska Pipeline Advance Could Threaten Mackenzie Valley Pipeline," *Natural Gas Week*, March, 7, 2005.

[28] National Energy Board (Canada), "National Energy Board Issues Certificate of Public Convenience and Necessity for Mackenzie Valley Pipeline," Press release, Calgary, AB, March 10, 2011; "Mackenzie Pipeline Needs Federal Cash: Insiders," CBC News, January 12, 2011.

[29] Denali-the Alaska Gas Pipeline, "Denali Discontinues Gas Pipeline Project," Press release, Anchorage, AK, May 17, 2011.

[30] Cassandra Sweet, "TransCanada Sticks to Alaska Gas Pipeline Project Even as Rival Bows Out," *Wall Street Journal*, May 17, 2011.

[31] For additional analysis of natural gas markets, see CRS Report R41543, *Global Natural Gas: A Growing Resource*, by Michael Ratner.

[32] U.S. Federal Energy Regulatory Commission, *Order Issuing Certificate And Approving Abandonment*, Empire Pipeline, Inc., Docket No. CP10-493-000, issued May 19, 2011.

[33] EIA's 2011 and 2010 Annual Energy Outlook.

[34] National Energy Board of Canada, "Understanding Canadian Shale Gas," November 2009; Advanced Resources International, Inc., *World Shale Gas Resources: An Initial Assessment of 14 Regions Outside the United States*, Arlington, VA, prepared for the U.S. Energy Information Administration, April 2011, Chapter I, *http://www.adv-res.com* /pdf/ARI %20EIA%20Intl%20Gas%20Shale%20APR%202011.pdf

[35] This restructuring has been somewhat complicated across TransCanada's system. For an initial perspective, see Samantha Santa Maria, "TransCanada Files Final 2011 Mainline Tolls; Raises Rates Again," Platts, April 29, 2011.

[36] For example, in 2003 testimony before Congress, the Federal Reserve Chairman stated "... high gas prices projected in the American distant futures market have made us a potential very large importer.... Access to world natural gas supplies will require a major expansion of LNG terminal import capacity." The Honorable Alan Greenspan, Chairman, U.S. Federal Reserve Board, "Natural Gas Supply and Demand Issues," Testimony before the House Energy and Commerce Committee, June 10, 2003.

[37] Peter Morton, "Canada, U.S. on Course for LNG Collision," *National Post*, February 16, 2007.

[38] U.S. Dept. of Energy, "Energy Department Approves Gulf Coast Exports of Liquefied Natural Gas," Press release, May 20, 2011.

[39] Some in industry argue that an increase in demand and higher prices in foreign markets for natural gas would increase production and not increase U.S. natural gas prices.

[40] Kinder Morgan is considering more than doubling its Trans Mountain Pipeline capacity and expanding west coast shipping facilities. Enbridge has proposed the new Northern Gateway project. Both projects would send oil from Alberta to the British Columbia coast. Enbridge, "Northern Gateway at a Glance," press release, 2011, *http://www.northerngateway* .ca/project-info/northern-gateway-at-a-glance.

In: U.S. Oil Sources
Editor: Natalie B. Kelleher
ISBN: 978-1-62100-415-8

Chapter 3

KEYSTONE XL PIPELINE PROJECT: KEY ISSUES

Paul W. Parfomak, Neelesh Nerurkar, Linda Luther and Adam Vann

SUMMARY

Canadian pipeline company TransCanada has filed an application with the U.S. Department of State to build the Keystone XL pipeline, which would transport crude oil from the oil sands region of Alberta, Canada, to refineries in the United States. Keystone XL would have the capacity to transport 830,000 barrels per day, delivering crude oil to the market hub at Cushing, OK, and further to points in Texas. The project is expected to cost more than $7.0 billion, of which at least $5.4 billion would be spent on the U.S. portion. TransCanada is planning to build a short additional pipeline so that oil from the Bakken formation in Montana and North Dakota can also be carried on the Keystone XL pipeline.

The construction of petroleum facilities connecting the United States with a foreign country requires a Presidential Permit from the State Department based on a determination of national interest. An element of that determination for the Keystone XL project is the preparation of an Environmental Impact Statement (EIS) pursuant to the National Environmental Policy Act. On April 16, 2010, the State Department's draft EIS (DEIS) for the Keystone XL project was released for comment to the general public and interested federal, state, and local agencies. In response to EPA's and other agency comments, the State Department issued a supplemental draft EIS (SDEIS) on April 15, 2011. The State Department will accept public comments on the SDEIS until June 6, 2011. Following issuance of a final EIS, the State Department will solicit additional public comment and host a public meeting before making a determination on granting a Presidential Permit. The State Department estimates that it will release a final EIS and final Record of Decision and National Interest Determination by the end of 2011. The North American-Made Energy Security Act (H.R. 1938) would direct the President to issue a final order granting or denying the Presidential Permit for the Keystone XL pipeline by November 1, 2011. Whatever the State Department's decision, legal challenges appear likely.

Opponents to the Keystone XL pipeline project, primarily environmental groups and affected communities along the route, object to the project principally on the grounds that it supports "dirty" Canadian oil sands development, that it could pose an environmental risk to groundwater, and that it promotes continued U.S. dependency on fossil fuels. Arguments criticizing the greenhouse gas emissions of oil sands production are based to some degree on the assumption that limiting pipeline capacity to U.S. markets may limit output from Canada's oil sands.

Proponents of the Keystone XL pipeline, including Canadian agencies and petroleum industry stakeholders, point to energy security and economic benefits, such as job creation. Some contend that the Keystone XL project secures growing Canadian oil supplies for the U.S. market, which could offset imports from other, less dependable foreign sources. They also claim that if oil sands output cannot flow to the United States, infrastructure to export it to Asia will develop. Further, having recently permitted the original Keystone pipeline, a similar pipeline project, the State Department could face a consistency challenge if it were to come to a different conclusion on similar environmental issues for the Keystone XL permit.

International pipeline projects like Keystone XL are not subject to the direct authority of Congress, but numerous Members of Congress have expressed support for, or opposition to, the pipeline proposal because of its potential environmental, energy security, and economic impacts. Congress may have an oversight role stemming from federal environmental statutes that govern the pipeline's application review process.

INTRODUCTION

In September 2008, TransCanada (a Canadian company) applied to the U.S. Department of State for a permit to cross the U.S.-Canada international border with the Keystone XL pipeline project. If constructed, the pipeline would carry crude oil produced from the oil sands region of Alberta, Canada, to U.S. Gulf Coast refineries. Because the pipeline would connect the United States with a foreign country, it requires a Presidential Permit issued by the State Department. Members of Congress have expressed support for the proposed pipeline's potential energy security and economic benefits while others have expressed reservations about its potential environmental impacts.[1] Though Congress has no direct role in permitting the pipeline's construction, it may have an oversight role stemming from federal environmental statutes that govern the pipeline's application review process. The North American-Made Energy Security Act (H.R. 1938) would direct the President to issue a final order granting or denying the Presidential Permit for the Keystone XL pipeline by November 1, 2011 (§ 3).

This report describes the Keystone XL pipeline proposal and the process required for federal approval. It summarizes key arguments for and against the pipeline put forth by the pipeline's developers, federal agencies, environmental groups, and other stakeholders. The report discusses potential consistency challenges faced by the State Department in reviewing the pipeline application given its recent prior approvals of similar pipeline projects. Finally, the report reviews the constitutional basis for the State Department's authority to issue a Presidential Permit, and opponents' possible challenges to this authority.

Pipeline Description and Status

The U.S. portion of the Keystone XL pipeline project, as proposed, would pass through Montana, South Dakota, Nebraska, Oklahoma, and Texas (**Figure 1**). The pipeline would consist of approximately 1,380 miles of 36-inch-diameter pipe and have the capacity to transport 830,000 barrels per day (bpd) of crude oil to the United States, delivering up to roughly 200,000 bpd to an existing oil terminal in Oklahoma with the remainder sent further to points in Texas.[2]

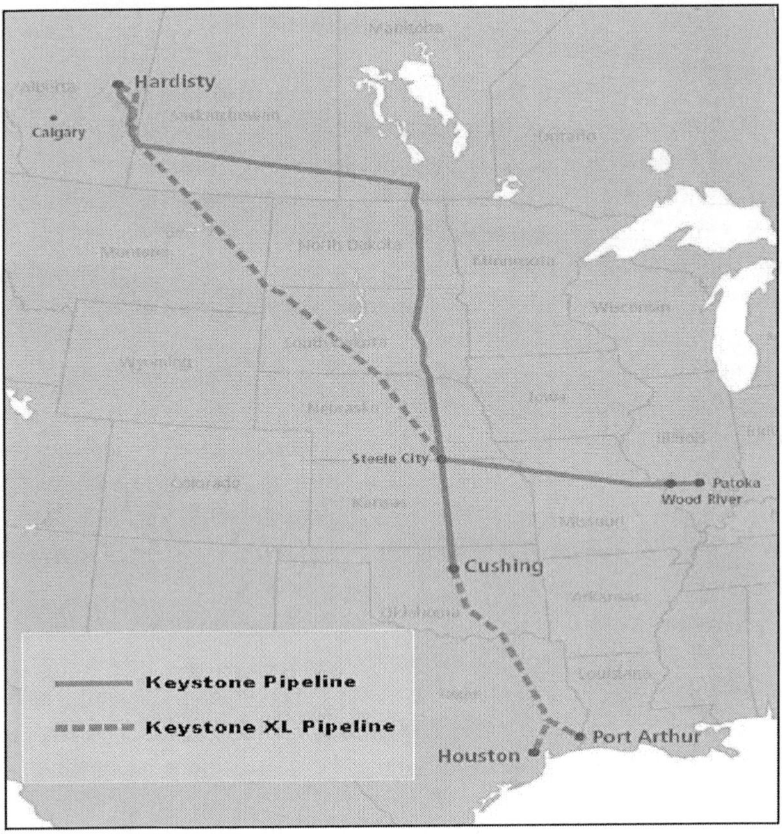

Source: TransCanada, Inc., *Keystone Pipeline System*, May 2010, http://www.transcanada.com/docs/Key_Projects/keystone_may_2010.pdf.

Note: Figure 1 shows the developer's "preferred alternative" for the Keystone XL pipeline route according to Presidential Permit application documents. For discussion of alternative routes, see the State Department EIS discussed below.

Figure 1. TransCanada Keystone Pipeline System Routes.

The Keystone XL project is expected to cost more than $7.0 billion, with the U.S. portion accounting for at least $5.4 billion of that total.[3] Current cost estimates include cost increases since the project's initial permit application was filed reportedly due to currency swings, changing regulatory requirements, and permitting delays.[4] The Keystone XL pipeline would be an extension of TransCanada's existing Keystone pipeline, which links the Alberta oil sands to refineries in Illinois and Oklahoma (**Figure 1**). The

Keystone pipeline received State Department approval on March 17, 2008, and began commercial operation in June 2010.

Keystone XL Extension to Bakken Oil Production

The Bakken formation is an unconventional oil resource in the Williston Basin, which underlies parts of North Dakota, eastern Montana, and northwestern South Dakota.[5] Current Bakken production is around 350,000 bpd, much of which is currently taken away by rail and truck, rather than by pipeline.[6] In part, this is because infrastructure has not kept up with rapid production growth in the Bakken region in recent years. Output is expected to increase significantly in the future, increasing the need for pipeline transportation capacity.[7]

TransCanada has signed contracts with Bakken oil producers to carry 65,000 bpd from the region via the Keystone XL pipeline. While not the full 100,000 bpd of capacity TransCanada had offered to oil producers, this was enough to justify adding the Bakken Marketlink Project, a pipeline running from Baker, MT, to the Keystone XL pipeline, which can then carry crude to the oil hub at Cushing, OK, and on to the Gulf Coast.[8] The Bakken Marketlink would have a 100,000 bpd capacity and is estimated to cost $140 million. It could start operating in 2013 if it and the Keystone XL pipeline receive regulatory approvals.[9]

These new Bakken contracts also improve the economics for Keystone XL, raising its committed capacity from 75% to near 90% of its projected 830,000 bpd.[10] Lower transportation costs and access to new markets may support investment in the Bakken. And TransCanada is not the only company adding pipeline capacity. Notably, Enbridge, another Canadian pipeline company, is building a 145,000 bpd pipeline to transport oil from the Bakken region to markets in the Midwest in the same time frame. According to Enbridge, sufficient pipeline capacity has been slow to emerge in the region because "they're smaller players in the Bakken. They are not able to make the 20-year commitments and it's been a lot of work to get them to commit to the level that [is] required to underwrite a major project out of the Bakken."[11]

Presidential Permit Application Requirements

Ordinarily, the U.S. government does not have permit authority for oil pipelines, even interstate pipelines. This is in contrast to interstate natural gas pipelines, which, under Section 7(c) of the Natural Gas Act, must obtain a "certificate of public convenience and necessity" from the Federal Energy Regulatory Commission (FERC).[12] As discussed later in this report, in the absence of federal government siting authority, any applicable state laws would establish the primary siting authority for oil pipelines. However, the construction, connection, operation, and maintenance of a pipeline that connects the United States with a foreign country requires executive permission conveyed through a Presidential Permit. Since the Keystone and proposed Keystone XL pipelines are designed for the importation of oil from Canada, their facilities require a Presidential Permit from the State Department.[13]

To issue a Presidential Permit, the State Department must find that issuance would serve the national interest.[14] In the course of processing such applications, the State Department consults extensively with concerned federal and state agencies, and invites public comment in arriving at its determination. With respect to the application submitted by TransCanada, the State Department concluded that the issuance of the Presidential Permit would constitute a major federal action that could have a significant impact upon the environment within the meaning of the National Environmental Policy Act (NEPA).[15] For this reason, the State Department must prepare an Environmental Impact Statement (EIS) to assess the environmental impacts that could result if the Keystone XL Pipeline project were approved.

Environmental Review under the National Environmental Policy Act

NEPA requires that federal agencies consider the environmental impacts of proposed actions and that they inform the public of those potential environmental impacts. In this case, NEPA applies to the Presidential Permit approval process. In processing Presidential Permit applications, the State Department must comply with NEPA.

Under the NEPA regulations followed by all federal agencies, an EIS must include a statement of the purpose and need for action, a description of all reasonable alternatives to meet that purpose and need, a description of the environment to be affected by those alternatives, and an analysis of the direct and indirect effects of the alternatives, including cumulative impacts.[16] Accordingly, the State Department must review and consider the potential environmental impacts of the entire pipeline, not just the facilities at the border crossing.[17] The EIS must also identify any state, tribal, or federal licenses, permits or approvals applicable to the project in the United States. In developing the EIS, the State Department must rely to some extent on information provided by TransCanada. For example, TransCanada's permit application included an Environmental Report which was intended to provide the State Department with sufficient information to understand the scope of potential environmental impacts of the project.[18]

The NEPA regulations require preparation of a draft EIS (DEIS) that is circulated for comment, followed by a final EIS that incorporates those comments.[19] During preparation of the DEIS, the State Department, in its role as "lead agency," requested input from any agency with jurisdiction by law with special expertise regarding any environmental impact associated with the project (referred to as "cooperating agencies").[20] Cooperating agencies for the Keystone XL project include the U.S. Environmental Protection Agency (EPA), the Department of Transportation's Pipeline and Hazardous Material Safety Administration (PHMSA), the Army Corps of Engineers, and the U.S. Department of Agriculture's Farm Service Agency, among others. In addition to its role as a cooperating agency in the EIS process, the EPA is also required to review the EIS itself to rate its adequacy and assess a project's environmental impacts.[21] EPA's rating may range from "lack of objections" to "environmentally unsatisfactory."

The State Department released its DEIS for the proposed Keystone XL Pipeline project for public comment on April 16, 2010.[22] The DEIS identified the developer's "preferred alternative" (**Figure 1**) for the project as well as other alternatives under consideration. The public comment period officially closed on July 2, 2010. On July 16, 2010, the EPA issued its rating of the DEIS, determining that it was "inadequate" and identifying a number of potential environmental impacts that had not been sufficiently addressed.[23] Among other concerns, EPA believed that the purpose and need of the project had been too narrowly crafted, that impacts to air and water quality were not fully analyzed, and that pipeline safety procedures were inadequate.

Following closure of the comment period, but before a final EIS was issued, Secretary Clinton made public statements that she was "inclined to" approve the pipeline. Environmental groups reportedly have argued that this statement indicated that a final decision had been made before the NEPA process was complete.[24] However, it is likely that these arguments were mooted by the State Department's supplementing the DEIS.

The State Department issued a supplemental draft EIS (SDEIS) on April 15, 2011. The SDEIS addresses comments, concerns, and recommendations from the public and various agencies, including EPA and PHMSA. It also incorporates new information that became available on the proposed project and on issues and resources related to the potential impacts of the proposed project since the DEIS was issued. The SDEIS is available for public comment from April 22, 2011, to June 6, 2011. After consideration of public and relevant agency comments, a final EIS may be issued. The final EIS will identify the chosen alternative. However, the NEPA review is not complete until issuance of a Record of Decision (ROD), formalizing the State Department's selection of an alternative.

A Record of Decision and National Interest Determination

It is important to note that, in arriving at its final decision regarding a Presidential Permit, NEPA requires only that the State Department assess the environmental consequences of an action and its alternatives before proceeding. If the adverse environmental effects of the proposed action are adequately identified and evaluated, the agency is not constrained by NEPA from deciding that other benefits outweigh the environmental costs and moving forward with the action. However, in addition to making its final decision under NEPA, the State Department must make a National Interest Determination under Executive Order 13337, which requires such a determination for pipeline projects that cross international borders. The Presidential Permit requires a determination that the proposal is in the national interest. It is possible that, based on environmental impacts, a project may be deemed not in the national interest. Under previous Presidential Permit applications for pipeline projects, the ROD and the National Interest Determination were issued in the same document.[25]

The State Department has announced that it will solicit public comment and host a public meeting in Washington, DC, before it makes a determination under Executive Order 13337 on whether issuance of this permit is in the U.S. national interest.[26] It is unclear how, or if, any public comments may be incorporated into a final ROD. The North American-Made Energy Security Act would direct the President to expedite the State Department's permit review process, requiring a final decision no later than November 1, 2011.

State Siting and Environmental Approvals

As noted above, the federal government does not exercise siting authority over oil pipelines. However, siting for the Keystone XL pipeline still must comply with any applicable state laws. These laws vary from state to state. South Dakota, for example, required TransCanada to apply for a permit for the Keystone XL pipeline from the state public utility commission, which issued the permit on April 25, 2010.[27] Montana requires a certificate from the state's

Department of Environmental Quality,[28] but has not yet granted one for the Keystone XL project. Nebraska does not appear to have any permitting requirements that apply specifically to the construction and operation of oil pipelines, although a state statute does include an "eminent domain" provision, which grants eminent domain authority to oil pipeline companies that are unable to obtain the necessary property rights from the relevant property owners.[29] A number of additional approvals and permits required by the states along the proposed route are summarized in TransCanada's Presidential Permit application.[30] All of the aforementioned state approvals are in various stages of review along the proposed Keystone XL pipeline route.

ARGUMENTS FOR AND AGAINST THE PIPELINE

Proponents of the Keystone XL pipeline, including Canadian agencies and U.S. and Canadian petroleum industry stakeholders, base their positions primarily on increasing the diversity of the U.S. petroleum supply and economic benefits to the United States, including job creation. Opponents, primarily environmental groups and affected communities along the route, object to the project principally on the grounds that Canadian oil sands development has negative environmental impacts and that it promotes continued U.S. dependency on fossil fuels. These issues are further discussed below.

Impact on U.S. Energy Security

In its Presidential Permit application, TransCanada asserts that constructing the proposed Keystone XL pipeline is in the U.S national interest to maintain adequate crude oil supplies for U.S. refineries. The application argues that the pipeline will allow U.S. refiners to substitute Canadian supply for other foreign crude supply and to obtain direct pipeline access to secure and growing Canadian crude output. In particular, the application asserts that the pipeline would allow the United States to decrease its dependence on foreign crude oil supplies from Mexico and Venezuela, the two largest oil importers into the U.S. Gulf Coast.[31] In its draft EIS for the project, the State Department similarly finds that the Keystone XL pipeline "would counteract insufficient domestic crude oil supply while reducing U.S. dependence on less reliable foreign oil sources."[32] These arguments have taken on additional weight in light of the ongoing political unrest in the Middle East, which has disrupted oil production in Libya, a significant oil exporter, and has caused a spike in global crude oil prices.

Canadian Oil Imports in the Overall U.S. Supply Context

Gross U.S. imports of crude oil and petroleum products averaged 11.8 million barrels per day (Mbpd) in 2010.[33] Exports averaged 2.3 Mbpd, leaving net imports at 9.4 Mbpd.[34] U.S. net imports declined each year between 2005 and 2010 as a result of lower total oil demand and higher domestic supply. Domestic demand has decreased by about 1.7 Mbpd versus 2005 levels due largely to the economic recession. Meanwhile, U.S. production of oil and oil alternatives (including crude oil, natural gas liquids, and biofuels) has increased by 1.4 Mbpd since 2005. As a result, net imports fell by roughly 3.1 Mbpd since 2005.[35] Some of this decline could be mitigated in the near term as oil demand recovers from the recession or if domestic supply were to fall. However, there is increasing consensus among forecasters that U.S. net oil imports have passed their high water mark already and may remain relatively flat in the long run.[36]

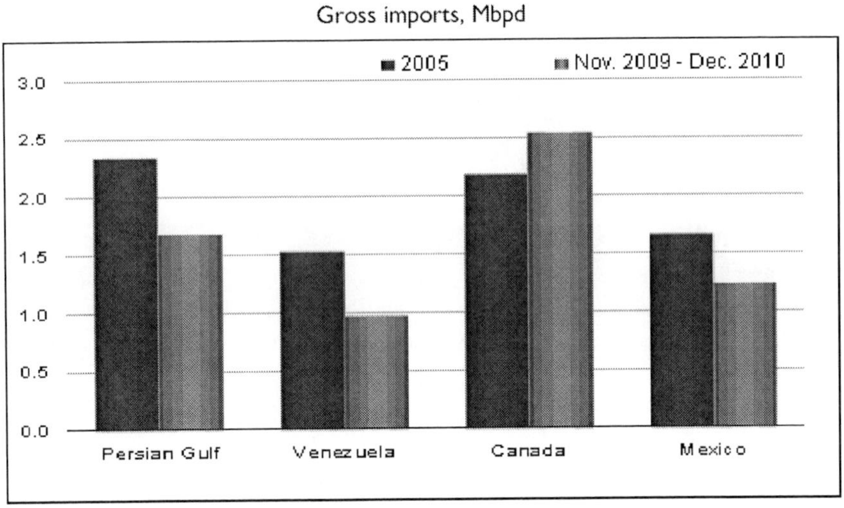

Source: U.S. Energy Information Administration, *"Petroleum Navigator: U.S. Imports by Country of Origin,"* December 12, 2010, http://www.eia.gov/dnav/pet/pet_move_impcus_a2_nus_ep00_im0_mbblpd_m.htm.

Figure 2. U.S. Oil Imports, Selected Sources.

Monthly imports in Mbpd on a 12-month moving average, Jan. 2000 to Oct. 2010

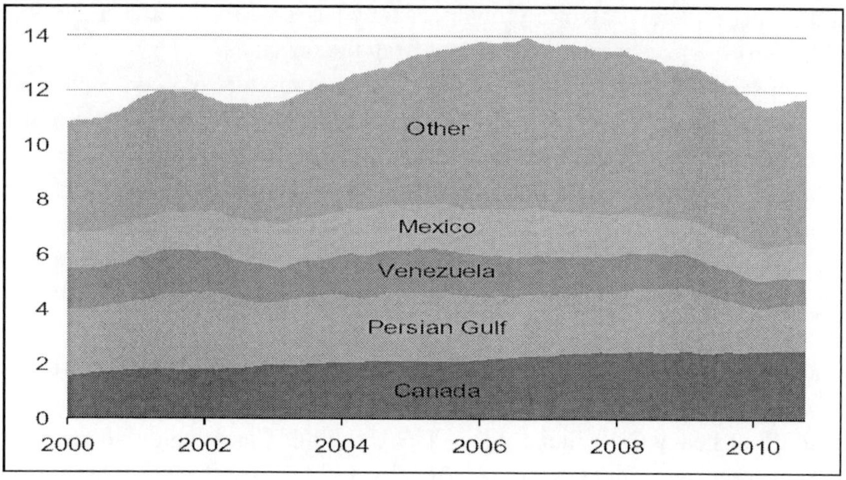

Source: U.S. Energy Information Administration, *"Petroleum Navigator: U.S. Imports by Country of Origin,"* December 12, 2010. http://www.eia.gov/dnav/pet/pet_move_impcus_a2_nus_ep00_im0_mbblpd_m.htm.

Figure 3. Total U.S. Oil Imports.

Among the largest sources of U.S. gross oil imports are Canada (2.5 Mbpd), the Persian Gulf (1.7 Mbpd), and Mexico (1.3 Mbpd). Imports from the latter two sources have decreased in recent years in part due to lower need for imports described above and in part due to developments in those countries (**Figure 2**). All major Persian Gulf exporters are members of the Organization of the Petroleum Exporting Countries (OPEC), which cut production in 2009 to support oil prices. Mexican production has been falling since 2004 because new oil developments have not been able to offset depletion at Mexico's giant Cantarell field. Imports from Venezuela, another key source of U.S. imports, has also fallen. Venezuelan production never fully recovered after a strike at its national oil company, *Petróleos de Venezuela*, in 2002-2003. Venezuelan production today is nearly 1 Mb/d less than that achieved in 2001. In recent years, Venezuela has also been trying to diversify business away from the United States, for example, by increasing exports to China.[37]

Meanwhile, Canadian production and exports to the United States have increased, primarily due to growing output from the oil sands in Western Canada. Energy markets in the United States and Canada are well integrated by pipeline infrastructure, and nearly all Canadian energy exports go to U.S. consumers. Canadian oil production has increased about 0.2 Mbpd since 2005, and exports to the United States increased by 0.4 Mbpd (**Figure 3**).[38] Canadian oil production is expected to grow by as much as 1.6 Mbpd between 2009 and 2025, mostly through increased output from the oil sands.[39]

Oil sands (also referred to as tar sands) are a mixture of clay, sand, water, and heavy black viscous oil known as bitumen. Oil sands require more processing than conventional crude oil. Oil sands are processed to extract the bitumen, which can then be upgraded into "syncrude" that is suitable for pipeline transport. Canada's oil sands production is expected to be exported as either a light, upgraded synthetic crude or a heavy crude oil that is a blend of bitumen diluted with lighter hydrocarbons to ease transport. The bulk of oil sands supply growth is expected to be in the form of the latter.[40] Most oil sands imports into the United States currently go to the Midwest, where some refineries are investing in complex refining capacity to process growing volumes of heavy Canadian crude.[41] The U.S. Gulf Coast region already has a large amount of complex refining capacity and is considered potentially well suited for processing Canadian heavy crude oil.[42] Gulf Coast refiners currently process heavy crudes from Venezuela, Mexico, and elsewhere. Complex refineries in the Gulf Cost may be best equipped to handle a large increase of heavy oil sands crude, though they may still need to adjust processes and make new capital investments in equipment to accommodate particular crudes' characteristics,[43] especially if the new Canadian crudes will be used in large amounts.[44] There are 15 refineries within Keystone XL's proposed delivery area in Texas that currently process heavy crude oil which is similar in composition to the oil that the pipeline would bring down.[45]

With expanded pipeline capacity extending to the U.S. Gulf Coast, Canadian oil sands crude may compete with other heavy crudes such as those from Mexico, Venezuela, and elsewhere.[46] It is difficult to predict precisely how this competition will play out, but it may take place through shifting discounts or premiums on crude oils from various sources.[47] It may be possible for Canadian oil supplies to effectively "push out" waterborne shipments from other countries, although this depends on a wide range of market conditions. Waterborne crudes may more easily go to other destinations than Canadian crudes, though like Canadian crudes they can be tied to specialized refining capacity, as is true for Venezuelan heavy crudes.

In 2010, 98% of Canada's oil exports went to the United States, mostly through north-south pipelines. One oil pipeline extends from Alberta to Canada's west coast, the Trans Mountain Pipeline, which is owned by Houston-based Kinder Morgan and has a capacity of 300,000 bpd. Some of the oil from the Trans Mountain Pipeline is loaded onto tankers and shipped from

Vancouver. Currently, about 90% of the crude shipped out by sea goes to California, with the remainder shipped to the U.S. Gulf Coast and Asia.[48]

There are proposals to increase the capacity for oil from Alberta to reach the Canadian west coast. Kinder Morgan is considering expanding the Trans Mountain Pipeline to 700,000 bpd, more than doubling its existing capacity, and expanding west coast shipping facilities. Enbridge has proposed a new pipeline: the Northern Gateway project would have a 525,000 bpd capacity to send oil from Edmonton to Kimat, British Columbia.[49] These projects reflect anticipated growth of Western Canadian oil production and an interest by Canadian oil producers to diversify their available markets beyond U.S. customers. Both proposals have received criticism from environmentalists. Because it would require construction of a completely new pipeline, Northern Gateway in particular has been criticized by some environmental and First Nations groups.[50]

If export capacity to the west coast is expanded it could increase the amount of Canadian crude oil going to non-U.S. markets. Canadian oil sales to Asian markets, where oil demand is growing rapidly, are more likely to develop if greater shipments to the United States are not possible.[51] A study commissioned by the Department of Energy concluded that:

> if pipeline projects to the BC [British Columbia] coast are built, they are likely to be utilized. This is because of the relatively short marine distances to major northeast Asia markets, future expected growth there in refining capacity and increasing ownership interests by Chinese companies especially in oil sands production. Such increased capacity would alter global crude trade patterns. Western Canadian Sedimentary Basin (WCSB) crudes would be "lost" from the USA, going instead to Asia. There they would displace the world's balancing crude oils, Middle Eastern and African predominantly OPEC grades, which would in turn move to the USA. The net effect would be substantially higher U.S. dependency on crude oils from those sources versus scenarios where capacity to move WCSB crudes to Asia was limited.[52]

Economic Impact of the Pipeline

In addition to supply diversity arguments, some Keystone XL pipeline proponents support the project based on economic benefits associated with expanding U.S. pipeline infrastructure. A recent study by the Energy Policy

Research Foundation, for example, concludes that "the Keystone expansion would provide net economic benefits from improved efficiencies in both the transportation and processing of crude oil of $100 million-$600 million annually, in addition to an immediate boost in construction employment."[53] A 2009 report from the Canadian Energy Research Institute (CERI) commissioned by the American Petroleum Institute similarly concludes that:

> As investment and production in oil sands ramps up in Canada, the pace of economic activity quickens and demand for US goods and services increase rapidly, resulting in an estimated 343 thousand new US jobs between 2011 and 2015. Demand for US goods and services continues to climb throughout the period, adding an estimated $34 billion to US GDP in 2015, $40.4 billion in 2020, and $42.2 billion in 2025.[54]

These CERI estimates apply to the entire oil sands industry, however, not only the Keystone XL project, and they are derived from a proprietary economic analysis which has not been subject to external review. Some stakeholders point to State Department and other studies reporting much lower anticipated economic benefits.[55] Consequently, it is difficult to determine what specific economic and employment impacts may ultimately be attributable to the Keystone XL pipeline. Nonetheless, given the physical scale of the project, it could be expected to increase employment and investment at least during construction.

Canadian Oil Sands Environmental Impacts

Oil production from oil sands is controversial because it has significant environmental impacts, including emissions of greenhouse gases during extraction and processing, disturbance of mined land, and impacts on wildlife and water quality.[56] Because bitumen in oil sands cannot be pumped from a conventional well, it must be mined, usually using strip mining or open pit techniques, or the oil can be extracted with underground heating methods.[57] Large amounts of water and natural gas are also required (for heating) during the extraction process.[58] The magnitude of the environmental impacts of oil sands production, in absolute terms and compared to conventional oil production, has been the subject of numerous, and sometimes conflicting, studies and policy papers.[59] Some stakeholders who object to oil sands projects oppose the Keystone XL pipeline because it expands access to new markets for the oil produced by those projects, thereby encouraging what they consider to be further environmentally destructive oil sands development. As discussed earlier, however, if oil sands production can be diverted to other markets (e.g., Asia), preventing the Keystone XL project may not necessarily limit oil sands development.[60]

Possible Risks to the Ogallala Aquifer

The proposed route of the Keystone XL pipeline passes across significant portions of the Ogallala Aquifer (**Figure 4**), one of the world's largest known aquifers and the primary source of groundwater for approximately 20% of U.S. agricultural production.[61] Because the aquifer is relatively close to the surface, some stakeholders are concerned that a release from the pipeline could potentially contaminate the aquifer with oil, jeopardizing its use for farming and drinking water and causing significant ecosystem damage. These concerns have been heightened in the wake of the 2010 spill from an Enbridge oil pipeline in Marshall, MI, which released 819,000 gallons of crude into a tributary of the Kalamazoo River. Furthermore, a recent report by the Natural Resources Defense Council (NRDC) argues that the Keystone XL pipeline could be more likely to fail and cause environment damage than other crude oil pipelines because the bitumen mixture it would carry is "significantly more corrosive to pipeline systems than conventional crude," among other reasons.[62] Canadian officials and other stakeholders have rejected these arguments, however, citing factual inaccuracies and a flawed methodology in the analysis, which compares pipeline spill rates in Canada to those in the United States.[63]

In its draft EIS for the Keystone XL pipeline project, the State Department states that "there is the possibility that a release could migrate through the overlying surface materials and enter a groundwater system."[64] Nonetheless, the department concludes that "the probability of a large spill occurring is very low, and, consequently, risk of environmental impacts is minimal."[65] The draft EIS views the risks of aquifer damage from the Keystone XL pipeline independently from such risks from other pipelines. Because the probability of a pipeline spill and subsequent groundwater contamination cannot be known with certainty, however, debate as to the groundwater risk potentially posed by the Keystone XL pipeline will likely continue.[66]

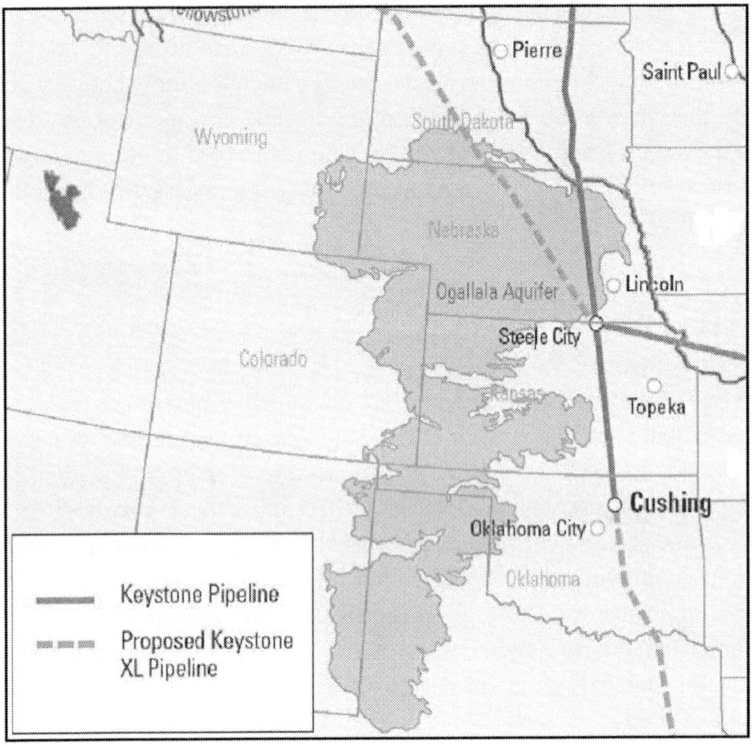

Source: Natural Resources Defense Council, *Say No to Tar Sands Pipeline*, November, 2010, p. 3.

Figure 4. Keystone XL Pipeline Route Across the Ogallala Aquifer.

Fossil Fuels Dependence

Some stakeholders object to the Keystone XL pipeline because it would increase U.S. supplies of oil, and thereby perpetuate the nation's dependence on imported fossil fuels and increase carbon emissions from the transportation sector.[67] Acknowledging this concern, in a public forum on October 20, 2010, Secretary of State Clinton reportedly remarked that "we're either going to be dependent on dirty oil from the [Persian] Gulf or dirty oil from Canada … until we can get our act together as a country and figure out that clean, renewable energy is in both our economic interests and the interests of our planet."[68] Critics of the State Department's draft EIS assert that the environmental review overlooks the pipeline project's overall impact on greenhouse gas emissions, for example, from the extraction and refining processes. However, others have argued that whether the Keystone XL Pipeline is constructed would have little bearing on greenhouse gas emissions as there are likely to be other export routes available for Canadian oil sands crude, and therefore, the same crude oils would still be transported and refined, albeit in different geographies (e.g., China).[69]

CONSISTENCY OF STATE DEPARTMENT REVIEW

In addition to the specific arguments surrounding the Keystone XL pipeline project summarized above, the State Department faces a consistency issue in reviewing the Presidential Permit application. As **Figure 1** and **Figure 4** show, the Keystone XL pipeline follows a similar route, starting in the Alberta oil sands and crossing the Ogallala aquifer, as the earlier Keystone pipeline, which the State Department approved. In 2009, the State Department also approved the Alberta Clipper pipeline, designed to carry crude oil from the Alberta tar sands region to Wisconsin. Because of its prior approvals of the Keystone and Alberta Clipper pipelines, it might be difficult for the State Department to reach different conclusions on certain environmental issues in its review of the Keystone XL pipeline, and reject the permit application on that basis. Doing so could create political, and potentially legal, challenges to either its earlier environmental review, or the current one. Some observers maintain that, in its ultimate decision whether to grant the Keystone XL pipeline a Presidential Permit, the State Department may, to some extent, be constrained by recent approvals of similar projects.

APPENDIX. PRESIDENTIAL PERMITTING AUTHORITY

The executive branch has exercised permitting authority over the construction and operation of "pipelines, conveyor belts, and similar facilities for the exportation or importation of petroleum, petroleum products" and other products at least since the promulgation of Executive Order 11423 in 1968.[70] Executive Order 13337 amended this authority and the procedures associated with the review, but did not substantially alter the exercise of authority or the delegation to the Secretary of State in E.O. 11423.[71] However, the source of the executive branch's permitting authority is not entirely clear from the text of these Executive Orders. Generally, powers exercised by the executive branch are authorized by legislation or are inherent presidential powers based in the Constitution. E.O. 11423 makes no mention of any authority, and E.O. 13337 refers only to the "Constitution and the Laws of the United States of America, including Section 301 of title 3, United States Code."[72] Section 301 simply provides that the President is empowered to delegate authority to the head of any department or agency of the executive branch.

The legitimacy of this permitting authority has been addressed by federal courts. In *Sisseton v. United States Department of State*, the plaintiff Tribes filed suit and asked the court to suspend or revoke the Presidential Permit issued under E.O. 13337 for the TransCanada Keystone Pipeline.[73] The U.S. District Court for the District of South Dakota found that the plaintiffs lacked standing because they would be unable to prove their injury could be redressed by a favorable decision.[74] The court determined that even if the plaintiff's injury could be redressed, "the President would be free to disregard the court's judgment," as the case concerned the President's "inherent Constitutional authority to conduct foreign policy," as opposed to statutory authority granted to the President by Congress.[75]

The court further found that even if the Tribes had standing, the issuance of the Presidential Permit was a presidential action, not an agency action subject to judicial review under the Administrative Procedure Act (APA).[76] The court stated that the authority to regulate the cross-border pipeline lies with either Congress or the President.[77] The court found that "Congress has failed to create a federal regulatory scheme for the construction of oil pipelines, and has delegated this authority to the states. Therefore, the President has the sole authority to allow oil pipeline border crossings under his inherent constitutional authority to conduct foreign affairs."[78] The President could delegate his permitting authority to the U.S. Department of State, but delegation did not transform the permit's issuance into an agency action reviewable under the APA.[79]

In *Sierra Club v. Clinton*,[80] the plaintiff Sierra Club challenged the Secretary of State's decision to issue a Presidential Permit authorizing the Alberta Clipper pipeline. Among the plaintiff's claims was an allegation that issuance of the permit was unconstitutional because the President had no authority to issue the permits referenced in E.O. 13337 (in this case, for the importation of crude oil from Canada via pipeline).[81] The defendant responded that the authority to issue Presidential Permits for these border-crossing facilities "does not derive from a delegation of congressional authority ... but rather from the President's constitutional authority over foreign affairs and his authority as Commander in Chief."[82] The U.S. District Court for the District of Minnesota agreed, noting that the defendant's assertion regarding the source of the President's authority has been "well recognized" in a series of Attorney General opinions, as well as a 2009 judicial opinion.[83] The court also noted that these permits had been issued many times before and that "Congress has not attempted to exercise any exclusive authority over the permitting process. Congress's inaction suggests that Congress has accepted the authority of the President to issue cross-border permits."[84] Based on the historical recognition of the President's authority to issue these permits and Congress's implied approval through inaction, the court found the Presidential Permit requirement for border facilities constitutional.

ACKNOWLEDGMENTS

The authors would like to acknowledge the contributions of Kristina Alexander and Vanessa Burrows to the content of this report.

End Notes

[1] See, for example, House Energy & Commerce Committee, Subcommittee on Energy and Power, Hearing on The American Energy Initiative: Discussion Draft of H.R. ____, the North American Made Energy Security Act of 2011, May 23, 2011; U.S. Senator Max Baucus, Letter to Secretary of State Hillary Rodham Clinton, September 10, 2010, http://baucus.senate.gov/?p=press_release&id=179; U.S. Representative Henry A. Waxman, Letter to Secretary of State Hillary Rodham Clinton, July 2, 2010, http://democrats.energycommerce.house.gov/documents/20100706/ State.070210.Clinton.Keystone.XL.pdf.

[2] U.S. Department of State, *Supplemental Draft Environmental Impact Statement for the Keystone XL Oil Pipeline Project*, April 15, 2011. p. 1-4. An initial capacity of 700,000 bpd may be raised to 830,000 bpd by increasing the pumping capacity. The Keystone XL project had applied to the Pipeline Hazardous Materials Safety Administration to operate at slightly higher pressure than permitted in standard regulations, which would have enabled a 900,000 bpd capacity, but it withdrew its applications for such a Special Permit in August, 2010.

[3] TransCanada Keystone Pipeline, L.P., Application of TransCanada Keystone Pipeline L.P. for a Presidential Permit Authorizing the Construction, Operation, and Maintenance of Pipeline Facilities for the Importation of Crude Oil to be Located at the United States-Canada Border, U.S. Dept. of State, September 19, 2008, p. 10, *http://www.keystonepipeline-xl.state*? OpenFileResource.

[4] "TransCanada Expects $1-Billion Cost Escalation for Keystone XL Pipeline," Canadian Press, February 17, 2011.

[5] Steven G. Grape, *Technology-Based Oil and Natural Gas Plays: Shale Shock! Could There Be Billions in the Bakken?*, Energy Information Administration, U.S. Department of Energy,

November 2006, http://www.eia.doe.gov/ pub/oil_gas/natural_gas/feature _articles/2006 /ngshock/ngshock.pdf.

[6] Nathan Vanderklippe, "TransCanada to move U.S. crude through Keystone," *The Globe and Mail*, January 26, 2011.

[7] Energy Information Administration, U.S. Department of Energy, *Annual Energy Outlook 2011 Early Release*, December 16, 2010, p. 8, http://www.eia.gov/forecasts/aeo/pdf/0383er(2011).pdf.

[8] Jeffrey Jones, "TransCanada plans U.S. Bakken pipeline link," *Reuters*, January 20, 2011.

[9] TransCanada, "TransCanada to Transport U.S. Crude Oil to Market Bakken Open Season a Success," press release, January 11, 2011, http://www.transcanada.com/5631.html.

[10] Vanderklippe, 2011.

[11] Lauren Krugel, "TransCanada attracts support for Montana-to-Oklahoma crude pipeline," *The Canadian Press*, January 20, 2011.

[12] 15 USC § 717f(c).

[13] See Executive Order 13337, "Issuance of Permits With Respect to Certain Energy-Related Facilities and Land Transportation Crossings on the International Boundaries of the United States," 69 *Federal Register* 25299, as amended, and Department of State Delegation of Authority No. 118-2 of January 26, 2006. The source of Permitting Authority for relevant Executive Orders is discussed further in the **Appendix**.

[14] Executive Order 13337, at Sec. 1(g).

[15] 42 U.S.C. § 4321 et seq.

[16] 40 C.F.R. Part 1502.

[17] Presidential Permits state that, by authorizing the permit, the State Department has considered requirements of Section 7 of the Endangered Species Act (16 U.S.C. 1536) and other statutes related to environmental concerns, the National Historic Preservation Act (NHPA) of 1966 (16 U.S.C. Section 470f), and Executive Order 12898 of February 11, 1994 (59 *Federal Register* 7629), concerning environmental justice.

[18] Documents submitted by TransCanada are available online at http://www.keystonepipeline-xl.state.gov/clientsite/ keystonexl.nsf?Open.

[19] For more analysis of NEPA requirements, see CRS Report RL33152, *The National Environmental Policy Act (NEPA): Background and Implementation*, by Linda Luther.

[20] 40 C.F.R. § 1508.5. Also, Executive Order 13337 directs the Secretary of State to refer an application for a Presidential Permit to other specifically identified federal departments and agencies on whether granting the application would be in the national interest.

[21] For more information on the EPA's role in the NEPA process, "Environmental Impact Statement (EIS) Rating System Criteria" at *http://www.epa.gov/complianc* e/nepa/comments/ratings.html.

[22] Documents prepared by the U.S. Department of State related to its NEPA requirements are available online at http://www.keystonepipeline-xl.state.gov/clientsite/keystonexl.nsf?Open.

[23] U.S. Environmental Protection Agency, Letter to the U.S. Department of State on the draft EIS for the Keystone XL project, July 16, 2010, *http://yosemite.epa.gov* /oeca/webeis.nsf/%28PDFView%29/20100126/$file/20100126.PDF.

[24] Josh Funk, *Washington Post,* "Groups Ask Clinton to Recuse Self on Pipeline Bid," November 4, 2010.

[25] U.S. Department of State, "Department of State Record of Decision and National Interest Determination, TransCanada Keystone Pipeline, LP Application for Presidential Permit," February 25, 2008.

[26] U.S. Department of State, "State Department Announces Next Steps in Keystone XL Pipeline Permit Process," press release, March 15, 2011, http://www.state.gov/r/pa/prs /ps/2011/03/158402.htm.

[27] South Dakota Public Utilities Commission, Final Decision and Order; Notice of Entry Before the Public Utilities Commission of the State of South Dakota, In the Matter of the Application by Transcanada Keystone Pipeline, LP for a Permit Under the South Dakota

Energy Conversion and Transmission Facilities Act to Construct the Keystone Pipeline Project, HP07-001, http://puc.sd.gov/commission/orders/HydrocarbonPipeline/2008/hp07-001.pdf.

[28] Montana Major Facility Siting Act, Title 75, Chapter 20.

[29] Nebraska Rev. Stat. § 57-1101.

[30] TransCanada Keystone, L.P., *Keystone XL Project: Preliminary Environmental Report*, September 2008, Table 7, http://www.keystonepipeline-xl.state.gov/clientsite/keystonexl.nsf/preliminaryenvironmentalreport.pdf? OpenFileResource.

[31] TransCanada Keystone Pipeline, L.P., September 19, 2008, pp. 6-8.

[32] U.S. Department of State, *Draft Environmental Impact Statement for the Keystone XL Oil Pipeline Project*, April 16, 2010, p. 4-2.

[33] All data in this section are from the U.S. Energy Information Administration's (EIA's) *Petroleum Navigator* (*http://www.eia.gov/dnav/pet/pet_move_impcus_a2_nus_ep00_im0_mbbl_m.htm*) and *International Energy Statistics* (*http://tonto.eia.doe.gov/cfapps/ipdbproject/IEDIndex3.cfm*).

[34] For context, the United States consumes roughly 19 Mbpd, more than 20% of the world's oil market.

Net imports are gross or total imports less total exports. This section will focus on gross imports, though it should be noted that among U.S. petroleum exports about 0.2 Mbpd of petroleum products go to Canada and 0.4 Mb/d to Mexico.

[35] These data are based on full year 2010 estimates provided by the EIA's *Short Term Energy Outlook* (STEO), http://www.eia.doe.gov/emeu/steo/pub/contents.html. The STEO provides a balance of U.S. supply and demand.

[36] For more analysis, see CRS Report R41765, *U.S. Oil Imports: Context and Considerations*, by Neelesh Nerurkar.

[37] U.S. Energy Information Administration, "Country Analysis Brief: Venezuela," February 2010, http://www.eia.doe.gov/emeu/cabs/Venezuela/Oil.html.

[38] As in the United States, Canadian consumption fell due to economic downturn. This allowed the increment in exports to be higher than the increment in production.

[39] Canadian Association of Petroleum Producers (CAPP), *Crude Oil: Forecast, Markets, and Pipelines*, June 2010, p. 2, http://www.capp.ca/getdoc.aspx?DocId=173003.

[40] CAPP, 2010, p. 7.

[41] CAPP, 2010, p. 13. According to CAPP, refineries adding capacity to process heavy oil in the Midwest include those in Roxana, IL; Whiting, IN, and Detroit, MI.

[42] CAPP, 2010, p. 14.

[43] Baker Hughes, *Planning Ahead for Effective Canadian Crude Processing*, Baker Petrolite White Paper, 2010, http://www.bakerhughes.com/assets/media/whitepapers/4c2a3c8ffa7e1c3c740001d/file/28271-canadian_crudeoil_update_whitepaper_06-10.pdf.pdf&fs=1497549.

[44] For a description of which units refineries may need to add (or have added) to be able to process more Canadian oil sands supply, see Praveen Gunaseelan and Christopher Buehler, "Changing US Crude Imports Are Driving Refinery Upgrades," *Oil and Gas Journal*, August 10, 2009.

[45] U.S. Department of State, April 15, 2011. p. 1-4.

[46] Center for Energy Economics and Bureau of Economic Geology, *Overview of the Alberta Oil Sands*, University of Texas at Austin, 2006, p. 16, *http://www.beg.utexas.edu*/energyecon/documents/overview_of_alberta_oil_sands.pdf.

[47] For more about the U.S. refining system, see CRS Report R41478, *The U.S. Oil Refining Industry: Background in Changing Markets and Fuel Policies*, by Anthony Andrews, Robert Pirog, and Molly F. Sherlock.

[48] Lucretia Cardenas, "Kinder Morgan Says Eyes Fall Open Season For Trans Mountain Pipeline Expansion," *Platts*, March 24, 2011.

[49] Enbridge, "Northern Gateway at a Glance," press release, 2011, http://www.northerngateway.ca/project-info/ northern-gateway-at-a-glance. The project would also include a pipeline to allow the import of 193,000 bpd of condensate, a light hydrocarbon that can be blended with bitumen to allow pipeline transport.

[50] Derrick Penner, "Opposition to Enbridge Northern Gateway pipeline grows," *Vancouver Sun*, December 2, 2010.

[51] Edward Welsch, "TransCanada: Oil Sands Exports Will Go to Asia if Blocked in U.S.," Dow Jones Newswires, June 30, 2010.

[52] EnSys Energy & Systems, Inc., *Keystone XL Assessment: Final Report*, Prepared for the U.S. Department of Energy, Office of Policy & International Affairs, December 23, 2010, p. 118.

[53] Energy Policy Research Foundation, Inc., The Value of the Canadian Oil Sands (….to the United States): An Assessment of the Keystone Proposal to Expand Oil Sands Shipments to Gulf Coast Refiners, Washington, DC, November 29, 2010, p. 2, *http://www* .eprinc.org/pdf/oilsandsvalue.pdf.

[54] Canadian Energy Research Institute, *The Impacts of Canadian Oil Sands Development on the United States' Economy, Final Report*, Calgary, Alberta, October 2009, p. vii.

[55] National Wildlife Federation, "TransCanada Exaggerating Jobs Claims for Keystone XL," November 9, 2010, http://www.dirtyoilsands.org/files/Keystone_XL_Jobs_11-09-10.pdf.

[56] For more analysis of oil sands and their environmental impacts, see CRS Report RL34258, *North American Oil Sands: History of Development, Prospects for the Future*, by Marc Humphries.

[57] U.S. Bureau of Land Management, "About Tar Sands," web page, January 11, 2011, http://ostseis.anl.gov/guide/ tarsands/index.cfm.

[58] Cecilia Jamasmie, "The Challenges and Potential of Canada's Oil Sands," *Mining*, September-October 2010, pp. 7-8.

[59] For an example of contrasting views, see IHS CERA Inc., *Oil Sands, Greenhouse Gases, and US Oil Supply, Getting the Numbers Right*, 2010; and Natural Resources Defense Council, "Setting the Record Straight: Lifecycle Emissions of Tar Sands," November 2010.

[60] For more analysis of oil sands and their environmental impacts, see CRS Report RL34258, *North American Oil Sands: History of Development, Prospects for the Future*, by Marc Humphries.

[61] Jane Braxton Little, "The Ogallala Aquifer: Saving a Vital U.S. Water Source," *Scientific American*, March 30, 2009.

[62] Anthony Swift, Susan Casey-Lefkowitz, and Elizabeth Shope, Tar Sands Pipelines Safety Risks, Natural Resources Defense Council, February 2011, p. 6.

[63] Canadian Energy Resources Conservation Board, "ERCB Addresses Statements in Natural Resources Defense Council Pipeline Safety Report," Press release, Calgary, Alberta, February 16, 2011.

[64] U.S. Department of State, *Draft Environmental Impact Statement for the Keystone XL Oil Pipeline Project*, Appendix P, "Risk Assessment," April 16, 2010. p. 4-6.

[65] Ibid. p. 6-1.

[66] For more analysis of pipeline safety issues, see CRS Report R41536, *Keeping America's Pipelines Safe and Secure: Key Issues for Congress*, by Paul W. Parfomak.

[67] See, for example: Natural Resources Defense Council, *Tar Sands Invasion: How Dirty and Expensive Oil from Canada Threatens America's New Energy Economy*, May 2010.

[68] Darren Goode, "Clinton Seems Poised to Approve TransCanada Pipeline," *The Hill*, October 20, 2010.

[69] EnSys Energy & Systems 2010, p. 116.

[70] *Providing for the performance of certain functions heretofore performed by the President with respect to certain facilities constructed and maintained on the borders of the United States*, 33 Fed. Reg. 11741, August 16, 1968.

[71] *Issuance of Permits With Respect to Certain Energy-Related Facilities and Land Transportation Crossings on the International Boundaries of the United States*, 69 Fed. Reg. 25299, May 5, 2004.
[72] Ibid.
[73] 659 F. Supp. 2d 1071, 1078 (D. S.D. 2009).
[74] Ibid. at 1078.
[75] Ibid. at 1078, 1078 n.5.
[76] See ibid. at 1080-81.
[77] Ibid. at 1081.
[78] Ibid.
[79] Ibid. at 1082.
[80] 689 F.Supp.2d 1147 (D. Minn. 2010).
[81] Ibid. at 1162.
[82] Ibid.
[83] Ibid. at 1163 (citing 38 U.S. Atty Gen. 162 (1935); 30 U.S. Op. Atty. Gen. 217 (1913); 24 U.S. Op. Atty. Gen. 100; and Natural Resources Defense Council (NRDC) v. U.S. Department of State, 658 F.Supp.2d 105, 109 (D.D.C. 2009)). The court in *NRDC* held that the State Department's issuance of a presidential permit under Executive Order 13337 was not subject to judicial review under the Administrative Procedure Act for abuse of discretion because "the issuance of presidential permits is ultimately a presidential action." 658 F. Supp. 2d at 109, 111-12. The court said that to allow judicial review of such decisions would raise separation of powers concerns. Ibid. at 111.
[84] Ibid.; see also Youngstown Sheet and Tube Co. v. Sawyer, 343 U.S. 579 (1952) (establishing a three-part test for analyzing the validity of presidential actions in relation to constitutional and congressional authority).

In: U.S. Oil Sources
Editor: Natalie B. Kelleher
ISBN: 978-1-62100-415-8

Chapter 4

MIDDLE EAST AND NORTH AFRICA UNREST: IMPLICATIONS FOR OIL AND NATURAL GAS MARKETS

Michael Ratner and Neelesh Nerurkar

SUMMARY

Political unrest in the Middle East and North Africa (MENA) has contributed to higher oil prices and added instability to energy markets. Supply disruptions and fears about the possible spread of unrest to major exporters have pushed prices higher. Even if the crisis abates, some risk premium may persist to the degree that market participants fear such an event could occur again.

Higher oil prices can negatively impact the economies of oil importing countries. The cost of oil is the primary determinant of gasoline prices and prices of other petroleum products; increased costs can be a burden on households and many businesses. Rising import costs for oil, natural gas and petroleum-based products can be a drag on economic growth by negatively affecting the trade balance. This may slow the current economic recovery, though it is not expected to derail it.

Many energy policy options to address vulnerability to disruptions and higher prices, such as what is taking place in MENA, are long-term in nature. It takes a long time for the energy sector to make material shifts, be they through renewables, efficiency, or increased domestic oil and gas production.

Short-term energy policy options (as opposed to the broader national security and diplomatic issues) are limited. Oil exporters with spare production capacity, particularly Saudi Arabia, may make short-term decisions to try to moderate prices by adjusting production levels, but their ability and willingness to do so are often based on internal decisions.

Part of the U.S. energy policy debate around recent unrest has focused on whether it is appropriate to release oil from the Strategic Petroleum Reserve (SPR). The government holds the SPR to mitigate the impacts of a "severe energy supply interruption." Proponents of using the SPR point out that there is a disruption to oil production in Libya and the resulting price increase negatively impacts the U.S. economic recovery. Critics question whether this is the appropriate time to release oil from the SPR or whether it should be saved to guard against larger future disruptions, and emphasize that the SPR has not traditionally been viewed as a device to manipulate prices.

INTRODUCTION

Recent unrest in the Middle East and North Africa (MENA) region has affected international energy markets and put upward pressure on oil prices.[1] MENA is home to some of the world's largest oil and natural gas producers and exporters. The region's exporters account for roughly 40% of oil and 20% of natural gas traded internationally. MENA countries also hold some of the largest reserves of oil and natural gas. (See the **Appendix** for energy profiles of the MENA countries.) This paper examines the consequences of the recent unrest on oil and natural gas markets, both in the short term and longer term.

SHORT-TERM PRICING PRESSURES

Energy prices can fluctuate based on current energy supply and demand conditions, expectations about future supply and demand, and financial market conditions. As a major source of the world's oil and natural gas production, MENA is critical to global energy markets. Current unrest there has shut down some energy production and raised uncertainties about future supply from the region. Some regional producers are seeking to reassure global energy markets, amid fears that unrest could spread to major producers or disrupt regional commerce. Oil is traded globally and oil prices are set in a globally integrated market under relatively short-term contracts. Consequently, what happens in any one part of the market—especially as significant a component as MENA—can affect the price of oil everywhere regardless of where the oil is produced or consumed. Natural gas markets have also been affected, but natural gas is not traded as widely as oil. Outside of North America and the United Kingdom, natural gas is traded mostly under long-term contract, and international market pricing is not as flexible as oil pricing. Consequently the effects of MENA unrest vary more by region, with the greatest impact currently being felt in Europe.

Unrest Contributes to Higher Oil Prices

The Brent crude oil price (a key international reference or benchmark) increased by nearly $20 per barrel from December 17, 2010, when protests started in Tunisia, to $113 per barrel on March 8, 2011, by which point unrest had spread through Egypt, Bahrain, Libya, and elsewhere.[2] Oil prices had already been increasing throughout the fourth quarter of 2010—due largely to the global economic recovery and the resulting increase in oil demand—but threats to oil supply from, and oil transportation through, MENA pushed them up further than might otherwise have been the case. The unrest has contributed to higher prices by causing some supply reductions and giving rise to worries that unrest may spread to the region's larger energy producers.

The MENA region includes some of the world's largest oil exporting countries and most important shipping chokepoints. The region produces 29.1 million barrels per day (Mb/d), a third of global oil supply, and its oil exports equal roughly 40% of the world's oil trade (see Table 1). Unrest in Egypt in January and February raised concerns of disruptions to oil and natural gas moving between the Red Sea and the Mediterranean, highlighting transit risks. Turmoil in Libya, a much bigger oil exporter than Egypt, and unrest in the Persian Gulf state of Bahrain increased concern in oil markets. As of March 2, 2011, estimates for how much oil production has been shut down in Libya varied, ranging from 0.8 Mb/d to 1.2 Mb/d.[3] (For context, Libya produced 1.8 Mb/d last year, and the global oil market is roughly 86 Mb/d.) Several U.S. companies are invested in Libya's energy production, including ConocoPhillips, Hess, and Occidental Petroleum; they are minority partners in their joint ventures with the Libyan National Oil Company.

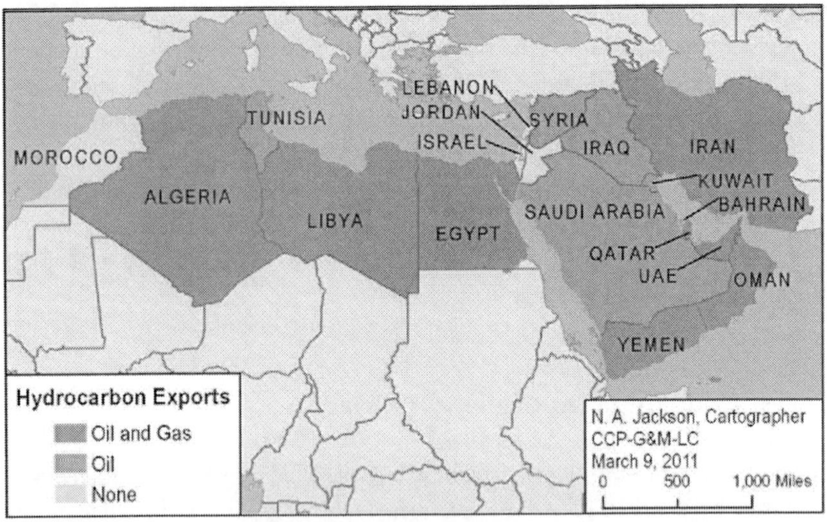

Source: Based on U.S. State Department information.

Notes: MENA includes Algeria, Bahrain, Egypt, Iran, Iraq, Israel, Jordan, Kuwait, Lebanon, Libya, Morocco, Oman, Qatar, Saudi Arabia, Syria, Tunisia, UAE, and Yemen. The unclear border between Egypt and Sudan is due to a border dispute.

Figure 1. Map of the Middle East and North Africa.

Saudi Arabia and Kuwait have publicly offered to make up any shortfall from the Libyan disruption by using their spare oil production capacity, and industry data show that Saudi Arabia has increased production by about 400,000 barrels per day. But some of the additional Saudi volumes may be of lower quality (heavier and higher in sulfur content) than the Libyan crudes, making them an imperfect substitute for certain refiners.[4] European refineries not equipped to process the lower quality crudes may have to source cargoes from places like Nigeria and Azerbaijan, which have crude closer in quality to the lost Libyan supplies.[5] Other refineries around the world, particularly some of the new refineries in Asia, may be better able to absorb the additional heavy, higher sulfur crude from Saudi Arabia, but at least in the short run there may be a dislocation that may raise prices, particularly for light, low sulfur crude, as global supply is redistributed. The transition may have some leeway as a number of European refiners are undergoing prescheduled maintenance in March, though the situation may tighten as they come back into operation in April.[6]

Table 1. MENA Oil Data 2009 Figures in millions of barrels a day (mb/d) or share of global totals

	Consumption	Production	% of World Production	Production Capacity	Exports	% of Global Oil Trade
Saudi Arabia	2.4	9.8	11.6%	12.8	7.3	13.8%
Iran	1.7	4.2	4.9%	4.3	2.5	4.7%
UAE	0.5	2.8	3.3%	3.1	2.3	4.4%
Iraq	0.6	2.4	2.8%	2.4	1.8	3.3%
Kuwait	0.4	2.5	3.0%	2.8	2.1	4.0%
Algeria	0.3	2.1	2.5%	2.1	1.8	3.4%
Libya	0.3	1.8	2.1%	1.8	1.5	2.9%
Qatar	0.1	1.2	1.4%	1.4	1.1	2.0%
Oman	0.1	0.8	1.0%	0.8	0.7	1.3%
Yemen	0.2	0.3	0.3%	0.3	0.1	0.2%
Syria	0.3	0.4	0.5%	0.4	0.1	0.2%
Other MENA	1.5	0.8	1.0%	0.8	[a]	NA
MENA Total	**8.4**	**29.1**	**34.5%**	**33.0**	**20.7**	**39.1%**
World Total	**84.4**	**84.4**	**100.0%**	**88.5**	**52.9**	**100.0%**

Source: Multiple U.S. Energy Administration databases and the BP Statistical Review of World Energy 2010.

Notes: Italics denotes members of the *Organization of the Petroleum Exporting Countries (OPEC)*. Production includes natural gas liquids.

a. These countries are net importers with about 0.7 Mb/d coming from other countries.

Of greater concern to oil markets is the risk that unrest may spread to other, larger producing countries in the region.[7] Of particular concern would be if unrest spreads to Saudi Arabia. It is the world's second-largest producer behind Russia. In addition to its current output, as of the end of 2010, Saudi Arabia also held about 80% of the world's spare oil production capacity[8]—capacity that is now being used to offset supply disruptions elsewhere. If significant geopolitical unrest emerges in Saudi Arabia, market concern may drive prices considerably higher. If any such unrest somehow disrupts Saudi oil production, it is difficult to predict how high oil prices may rise. The security of the Strait of Hormuz, between Iran and Oman, is also of great concern. A third of the world's oil trade passes through the Strait of Hormuz, and any closing or confining of transit could also substantially raise oil prices.

Impacts on the U.S. Oil Market

As of 2010, the United States imported half of the oil it consumed from abroad—9.4 Mb/d, of which 2.3 Mb/d came from MENA. More than 2.0 Mb/d came from three countries: Saudi Arabia, Algeria, and Iraq. Imports from Libya were less than 0.1 Mb/d; most of Libya's oil and other petroleum product shipments go to Europe. Because the oil market is globally integrated, the disruption of Libyan exports to Europe has affected the price of oil worldwide, including all U.S. imports and domestic production.

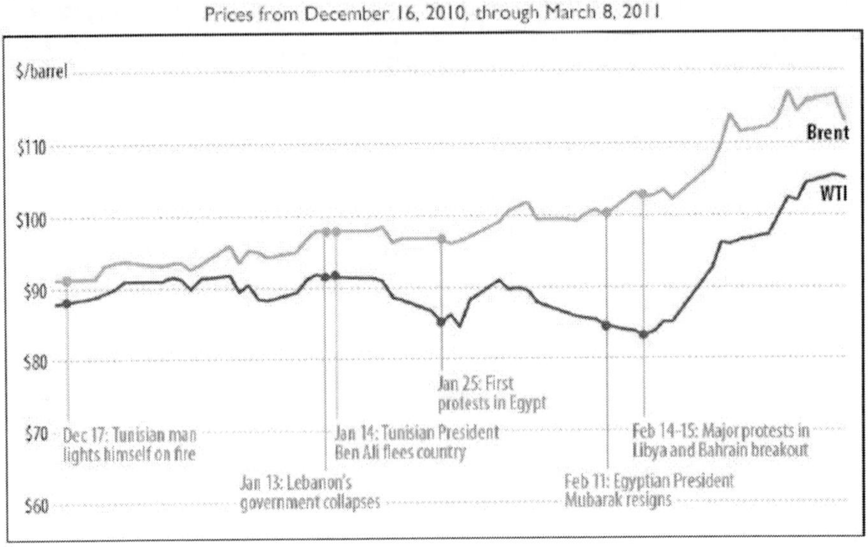

Source: U.S. Energy Information Administration - http://www.eia.doe.gov/dnav/pet/pet_pri_spt_s1_d.htm.

Notes: WTI is the main U.S. benchmark for crude oil prices, while Brent is the main European benchmark price. The difference between WTI and Brent prices is attributed to other supply and demand issues besides the MENA unrest. For more details, see footnote 2.

Figure 2. Recent Benchmark Oil Prices.

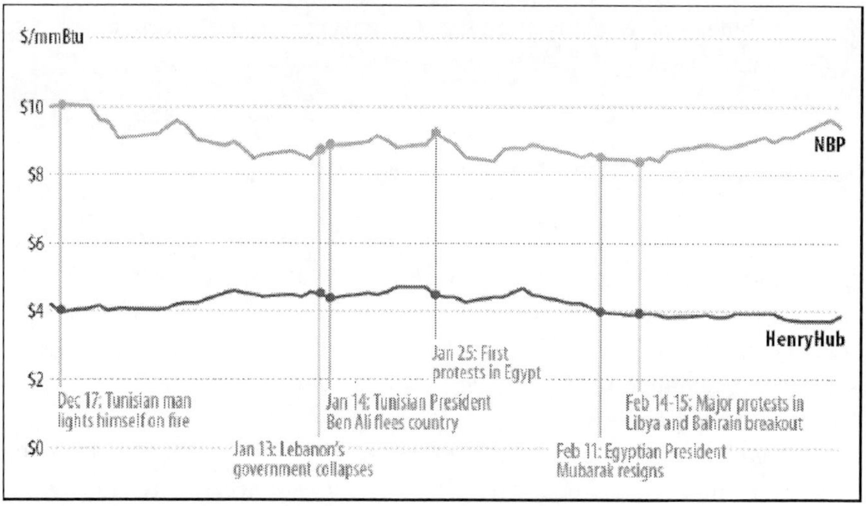

Data Source: U.S. Energy Information Administration, http://www.eia.doe.gov/dnav/ng/ng_pri_fut_s1_d.htm.

Notes: mmBtu = million British thermal units. NBP, the National Balancing Point, is a key trading point for natural gas in the United Kingdom and a reference price for the rest of Europe. Henry Hub is the equivalent for the United States and is the basis for the NYMEX price. There is no daily benchmark price for Asia, which tends to pay the highest prices for natural gas.

Figure 3. Recent Natural Gas Prices.

Economically, higher oil prices have a negative impact on household budgets and national economic growth for countries that are major importers.[9] According to recent testimony from Federal Reserve Chairman Ben Bernanke, while rising oil prices increase inflation and take income out of the pockets of consumers, reducing consumer spending and confidence, the recent increase does not yet appear to be a significant risk for the recovery or inflation.[10] Some analysts estimate that a sustained $10 increase in the oil price can reduce U.S. economic growth by roughly two-tenths of a percentage point.[11] However, this may not account for potential secondary effects such as through consumer confidence or other areas. The benchmark U.S. oil price on March 8, 2011, was $105 per barrel,[12] versus an average of $80 per barrel in 2010.[13]

Even under a scenario in which the United States produced as much oil as it consumes, international events would still affect U.S. oil prices as long as international trade were permitted. The drag on economic growth via the trade deficit would be muted in such a scenario. Households and businesses could still face higher (or lower) costs due to international events, but wealth would be redistributed domestically from energy consumers to producers to a greater degree than it is today.

U.S. Natural Gas More Insulated from MENA Turmoil

There is much less risk to the United States from MENA-related disruptions of natural gas supply. The United States imports about 16% of the natural gas it consumes, nearly all via pipeline from Canada. In 2009, MENA liquefied natural gas (LNG)[14] accounted for less than 1% of U.S. consumption. Since the beginning of the unrest, U.S. natural gas prices have actually declined.[15] (See **Figure 3**.)

The global natural gas market is much more regionally segmented than the oil market, and events in one part of the world do not necessarily affect prices everywhere. Only about 30% of natural gas is traded internationally, versus 60% for oil. Most natural gas exports from MENA go to Europe or Asia, where supplies are mostly sold under long-term contracts, insulating consumers in the short term from any price fluctuations caused by unrest. **Figure 3** highlights the minimal influence the unrest has had on natural gas markets. Some companies have had to scramble for LNG cargoes to limit their risk of a natural gas supply disruption, but there is ample spare capacity to meet this demand. The northern hemisphere winter is the main consumption season for natural gas, used primarily for heating, and prices will mostly be affected by changes in temperature.

Europe is the region most vulnerable to a natural gas shutdown in MENA, particularly from North African countries. The complete cut off by Libya of its natural gas exports, which almost exclusively go to Italy, has contributed to a 12% increase in European spot prices since major protests began. Russia has increased its pipeline exports to Italy to compensate for the shortfall. However, since a market vendor in Tunisia lit himself on fire in December, European prices are down almost 7%. If supply from Algeria—Europe's third-largest supplier of natural gas behind Russia and Norway—were for some reason reduced or cut off, Europe would be hard-pressed to replace this gas. (See the Algeria synopsis in the **Appendix** for additional information.) Europe's network of natural gas pipelines is designed to move natural gas from east to west, and requires more interconnections to move supplies throughout the region. In contrast, Asian prices have remained almost flat since the unrest started in Libya.

Table 2. MENA Gas Data 2009 figures in billion cubic feet (bcf) per year or % of global totals

	Consumption	Production	% of World Production	Exports	% of World Natural Gas Trade
Qatar	745	3,154	3.0%	2,408	7.7%
Algeria	1,016	2,876	2.7%	1,860	6.0%
Egypt	1,567	2,214	2.1%	647	2.1%
Oman	520	875	0.8%	408	1.3%
Libya	212	562	0.5%	349	1.1%
UAE	2,086	1,725	1.6%	248	0.8%
Iran	4,649	4,632	4.4%	200	0.6%
Yemen	4	18	<0.5%	15	<0.5%
Other MENA	4,359	4,072	3.8%	0	0%
MENA Total	**15,158**	**20,127**	**18.9%**	**6,135**	**19.7%**
World Total	**106,764**	**106,471**	**100.0%**	**31,184**	**100.0%**

Source: U.S. Energy Information Administration, multiple databases.

Notes: Italics denotes member of the *Gas Exporting Countries Forum*, a nascent cartel-like organization. Iran, Oman, and the UAE import natural gas in addition to exporting it.

LONG-TERM CONSIDERATIONS: RISKS MAY PERSIST

Oil and natural gas exporting countries in MENA rely heavily on revenues from energy exports to fund government budgets and fuel their economies. And MENA exports will make up a substantial component of world oil and gas trade for the foreseeable future. MENA countries hold approximately 61% of oil reserves and 45% of natural gas reserves. Changes in government policies that affect the development of these resources will have a long-term impact on global markets. MENA is home to 8 of the 12 countries in the Organization of the Petroleum Exporting Countries (OPEC)[16] and 5 of the 11 members of the Gas Exporting Countries Forum (GECF).[17]

If unrest persists in MENA countries, the long-term effects may be incorporated into oil and natural gas market expectations. Energy markets may evaluate the risk associated with exports from MENA differently in the future. The outbreak and persistence of unrest across the region has highlighted geopolitical risks that may not have been fully recognized before the crisis. Even after the current outbreak of unrest subsides, concerns that it could re-emerge may mean that some part of the current risk premium may persist.

Countries that undergo regime change may re-evaluate national energy development, production, and export policies. In the short term, regime change is likely to support relatively higher prices by creating uncertainty. Long-run risks from regime change and energy policy re-evaluation from energy exporting countries can vary on a case-by-case basis. Risks from regime change in countries such as Libya or Iran, historically sources of instability to the oil market, may be different than regime change in Saudi Arabia. On the other hand, market observers may judge that the revenue requirements of many regional producers create strong incentives for energy policy continuity in producing states, regardless of the makeup of their governments.

Even absent regime change, the current crisis may foster changes in MENA energy policy. Some governments have already increased public expenditures in apparent hope that it may dampen unrest, and these expenditures may increase revenue needs over the long term.[18] Increasing energy consumption in some exporting countries also creates complex choices for government and industry. Should increases in government expenditures be sustained, these countries may require greater revenues from oil and natural gas exports. Whether they attempt to do this through higher export volumes than would otherwise be the case (which may contribute to lower prices), or coordinating lower exports than would otherwise be the case to raise prices, is unclear.

POLICY CONSIDERATIONS

Policy options involving action or oversight from Members of the 112th Congress fall into three categories: short-term energy policies, long-term energy policies, and addressing the underlying political unrest. The oil market is global; no one country controls it, though the policies of larger producer and consumer nations can have more influence than others. Policy measures aimed at reducing the price of oil or U.S. import dependence tend to focus on increasing oil supply (or the supply of oil alternatives) and reducing demand.

But the energy industry has long lead times and is capital intensive, so it takes long periods of time to implement most supply and demand measures. As a result, there are limited short-term policy options, which Congress recognized in its justification for establishing the Strategic Petroleum Reserve (SPR). The SPR created a short-term policy tool that reduces the vulnerability of the United States to the consequences of a severe energy supply interruption. However, debates have continued concerning the appropriateness of using the SPR and what qualifies as "severe."

Much of the discussion around energy policy responses to MENA unrest is focused on oil. U.S. natural gas consumption, as mentioned above, is mainly supplied by domestic resources with some imports largely from Canada. As natural gas is currently not a substitute for oil, particularly in the transportation sector, policies to increase production will have minimal if any effect on U.S. oil consumption. There are proposals to encourage more natural gas use in the transportation sector, but these proposals are long-term in nature. The United States uses very little oil in electricity generation, making the effect of fuel switching to natural gas in that sector minimal. Globally, some U.S. companies have applied for permits to export natural gas, but the limited LNG additions would unlikely change the overall global natural gas picture.

Strategic Petroleum Reserve

The U.S. government holds 726.5 million barrels of crude oil in the SPR—equivalent to 75 days of imports—to respond to supply interruptions.[19] Oil from the SPR could reach markets within 13 days of the President authorizing its release at a maximum rate of 4.4 Mb/d, though any market impact could be more immediate once the market knows it has been made available. Further, some of the light, sweet barrels available in the SPR are a closer match to shut-in Libyan crude (or the Nigerian, Azeri, etc., crude that may substitute for it) than some of the heavier and/or higher sulfur crude that may come out of Saudi Arabia. An SPR release can be coordinated with other members of the International Energy Agency, which also hold strategic reserves.[20] However, using the SPR at any given point may leave less oil in storage to deal with subsequent disruptions.[21]

The Energy Policy and Conservation Act of 1975 (P.L. 94-163, EPCA), establishing the SPR, authorizes a drawdown upon a finding by the President that there is a "severe energy supply interruption"—an energy supply shortage that is likely to be of significant scope and duration and may cause major adverse impact on national safety or the national economy which results from an interruption in the supply of imported or domestic petroleum.[22] What qualifies as a severe interruption is a matter of debate (see below). EPCA also includes provisions for test sales and exchanges of crude in the SPR.[23]

In its FY2012 Budget Request, the Department of Energy (DOE) called for a sale of approximately 6 million barrels of oil for operational purposes. One of the 62 caverns that hold the reserve, one at the Choctaw Bayou SPR site in Louisiana that had originally been assessed as able to hold 7.5 million barrels, has leached toward the edge of the salt dome it sits under and may now only be able to safely hold 3.2 million barrels.[24] The excess oil is being moved to other locations, but the DOE is concerned about overfilling these other locations and has proposed selling the excess crude rather than risk losing it through leaks.[25] The sale will reduce the crude oil stocks to 74 days of imports.

Congressional views on the use of the SPR vary. Some have supported the sale of 6 million barrels for operational reasons, citing fiscal and environmental concerns about losing the oil.[26] Others have asked the Administration to stand ready to draw down the SPR if the situation in MENA deteriorates any further, posing greater risk to oil supply.[27] And still others are calling for an immediate SPR release to bring down prices.[28] Critics of tapping the reserve have questioned the Administration's motive in selling crude for operational reasons, and suggested that the SPR should be preserved for more severe disruptions, resisting calls that it be used to affect prices. Industry analysts disagree whether selling crude from the SPR will have a significant market impact. On March 8, 2011, officials stated that the Administration is monitoring the situation and that drawing down the SPR is an option under consideration.[29]

Other Policy Options

The SPR was established at least in part because there are few other short-term options to mitigate the effect of severe oil market disruptions. Price controls implemented in the 1970s, also responses to MENA oil supply disruptions, were ultimately rejected as counterproductive to long-term energy goals.[30] When prices were rising rapidly in 2008, Congress considered but did not pass a temporary "gas tax holiday." The proposal, S. 2890 in the 110th Congress, would have amended the Internal Revenue Code of 1986 to reduce the 18.4¢ per gallon federal gas tax and 24.4¢ diesel tax to zero during the summer driving season. It met significant opposition on concerns it would contribute to higher oil import dependence by spurring increased consumption and jeopardize programs funded by the Highway Trust Fund, the recipient of the fuel excise tax.[31]

Apart from the SPR, much of the policy debate stemming from the unrest in MENA and its impacts on oil markets has focused on options that take a long time to implement. Some Members of Congress and others cite the current unrest in MENA and the resulting increase in oil costs in backing proposals ranging from increasing support for domestic oil and gas production, and renewable energy, or mandating greater efficiency. Such proposals may contribute to lower import dependence but only over time, as they take months or years to make material impacts on liquid fuels supply and demand. Further, while such moves would reduce the drag on economic growth when oil prices rise, American consumers could still face higher gasoline prices at the pump due to international disruptions of oil supply, since oil prices are set in a global marketplace. Discussing these numerous options falls beyond the scope of this report.

Also, Congress may seek to use its authorization, appropriations, and oversight powers to monitor and affect the executive branch's efforts to address the underlying political unrest in MENA. For more information on political unrest in MENA, see the CRS Issue in Focus page on the Middle East and its associated reports.

APPENDIX. COUNTRY FACTORS

Political profiles of MENA countries are available via the CRS Issue in Focus page on the Middle East.

Algeria

Oil Exports: Algeria is in the top 15 of oil exporters, sending approximately 1.2 Mb/d around the world. Europe is the primary recipient of Algerian oil. Algeria is also a main exporter to Europe of naphtha, which is used to make high octane gasoline and jet fuel, and condensate, a light natural gas liquid.

Natural Gas Exports: Europe, mainly Italy, is Algeria's primary customer for natural gas. Algeria is Europe's third-largest supplier of natural gas, behind Russia and Norway, exporting by both pipeline and LNG. Much of Algeria's natural gas exports come from the Hassi R'Mel field. There are two main export pipelines—the Transmed pipeline going to Italy via Tunisia and the Maghreb-Europe gas pipeline going to Spain via Morocco—and two LNG export terminals, Arzew and Skikda. Arzew was the first LNG facility built in the world in 1964. There are three additional pipelines at various stages of development: the Medgaz pipeline going directly to Spain, the Galsi pipeline going directly to Italy, and the Trans-Sahara pipeline that would bring Nigerian natural gas to Europe via Algeria. Medgaz is scheduled to open in 2011, Galsi in 2012, and there is no date for the Trans-Sahara as it is unlikely to be constructed.

Supplemental Energy Concerns: None.

Bahrain

Oil Exports: Bahrain shares the Abu Safah field with Saudi Arabia and is a minor exporter of oil.

Natural Gas Exports: Bahrain is not an exporter of natural gas.

Supplemental Energy Concerns: Bahrain is linked by a causeway to the Eastern Province of Saudi Arabia, which serves as the production and export hub for most of Saudi Arabia's oil. As such, unrest in Bahrain may concern oil markets as a possible indication of potential risk to supplies from Saudi Arabia. Bahrain also hosts the headquarters of the U.S. Navy's Fifth Fleet, which plays an important role in providing maritime security in and around the Persian Gulf.

Egypt[32]

Oil Exports: Egypt is a minor exporter of oil.

Natural Gas Exports: Egypt is the 12th-largest exporter of natural gas, with cargoes going to the United States, Europe, and Asia. However, its main influence is its regional natural gas exports by pipeline to Israel and to Jordan, Lebanon, and Syria by the Arab Gas Pipeline. In 2009, Egypt supplied all the natural gas consumed in Jordan and Lebanon, half of Israel's consumption, and 13% of Syria's consumption.[33]

Supplemental Energy Concerns: Egypt's control of the Suez Canal and the Suez-Mediterranean (SUMED) oil pipeline raised concern during the recent unrest. If both of these transit options were closed it would likely take a few weeks for the oil and natural gas industry to find alternative routes. There would be additional costs to these alternative routes, and many cargoes would have to factor in the required time and cost of going around South Africa's Cape of Good Hope. There are plans to extend the Arab Gas Pipeline to Turkey in order to send natural gas to Europe.

Iran

Oil Exports: In 2009, Iran exported 2.2 Mb/d, making it the third-largest exporter, behind Saudi Arabia and Russia. Although Iran is a major exporter of oil, it imports refined products and has experienced shortages of some products, including gasoline. Kharg Island is Iran's main oil export terminal, with a loading capacity of 5 Mb/d.

Natural Gas Exports: Although Iran is the second-largest reserve holder of natural gas, it is a net importer. Iran exports small amounts of natural gas to Turkey, but imports from Turkmenistan. Additionally, Iran co-operates the South Pars/North Field, the world's largest natural gas field, with Qatar.

Supplemental Energy Concerns: Iran, along with Oman, is one of the littoral countries of the Strait of Hormuz, through which transits approximately 15.5 Mb/d, or nearly a third, of the world's oil trade. If this passage were threatened or closed, the impact on world oil markets would be tremendous, with prices likely reaching new highs. Iran is also a regional exporter of electricity. Iran's belligerent relationship with Israel adds instability to the marketplace, as an attack by either would have a big impact on oil exports and possibly natural gas exports if it extended beyond Iran's borders.

Iraq

Oil Exports: Iraq is one of the countries that could have tremendous growth in its oil production should the country remain stable. In 2009, Iraq exported almost 1.8 Mb/d, making it the 10th-largest exporter. The Basrah oil terminal has a 1.3 Mb/d capacity and is Iraq's main export terminal. Iraq also exports oil by pipeline to Turkey.

Natural Gas Exports: At present, Iraq does not export natural gas, although it does have substantial reserves.

Supplemental Energy Concerns: Iraq has the potential to be among the leading sources of global oil supply growth. It is home to the world's fourth-largest proven oil reserves, behind Saudi Arabia, Canada, and Iran, but development plans may be threatened by unrest.

Israel

Oil Exports: Israel is not an exporter of oil.

Natural Gas Exports: Israel is not an exporter of natural gas, but recent discoveries could make Israel self-sufficient in natural gas and possibly lead to exports.[34]

Supplemental Energy Concerns: Israel's influence on oil and natural gas markets comes from its strained relationships with some MENA countries, and its hostile relationship with Iran. The possibility of Israel attacking Iran over its nuclear program adds instability to the market.

Jordan

Oil Exports: Jordan is not an exporter of oil.

Natural Gas Exports: Jordan is not an exporter of natural gas.

Supplemental Energy Concerns: Jordan is a transit point on the Arab Gas Pipeline from Egypt, which also supplies natural gas to Lebanon and Syria. Rising energy prices were creating significant fiscal challenges for the Jordanian government prior to the recent unrest. Sustained high prices may complicate government efforts to respond to protestor demands and maintain a healthy fiscal balance.

Kuwait

Oil Exports: In 2009, Kuwait ranked as the 11th-largest exporter of oil. Given its proximity to Saudi Arabia, unrest in Kuwait may have a greater market impact than its relative importance as a producer would suggest.

Natural Gas Exports: Despite having ample natural gas reserves, Kuwait actually imports natural gas. In the future, Kuwait reportedly intends to increase its natural gas production to meet domestic demand.

Supplemental Energy Concerns: Kuwait shares an area referred to as the Partitioned Neutral Zone (PNZ) with Saudi Arabia. The resources of the PNZ are developed in a 50-50 relationship with Saudi Arabia.

Lebanon

Oil Exports: Lebanon is not an exporter of oil.

Natural Gas Exports: Lebanon is not an exporter of natural gas.

Supplemental Energy Concerns: Lebanon is a transit point on the Arab Gas Pipeline from Egypt.

Libya

Oil Exports: Prior to the ongoing crisis, Libya was exporting over 1.2 Mb/d of oil, which is approximately 3% of global exports. In addition to its exports, Libya historically has played an important role in the oil industry, with Muammar al Qadhafi and his supporters leading the nationalization effort of most of most Arab countries' oil industry assets in the early 1970s.

Natural Gas Exports: Libya's exports almost exclusively go to Italy and in 2009 made up almost 13% of that country's natural gas consumption. The Greenstream natural gas pipeline to Italy has ceased flows. Libya also exports a small amount of LNG.

Supplemental Energy Concerns: None.

Morocco

Oil Exports: Morocco is not an exporter of oil.

Natural Gas Exports: Morocco is not an exporter of natural gas.

Supplemental Energy Concerns: The Maghreb Europe Gas pipeline from Algeria to Spain transits Morocco.

Oman

Oil Exports: Oman is in the top 20 of oil exporters, accounting for about 1% of global exports.

Natural Gas Exports: Oman is a net exporter of natural gas, but does import a small amount from Qatar. Most of Oman's LNG goes to Asia.

Supplemental Energy Concerns: Oman, along with Iran, is one of the littoral countries of the Strait of Hormuz, through which transits approximately 15.5 Mb/d, or nearly a third, of the world's oil trade. If this passage were threatened or closed, the impact on world oil markets would be tremendous, with prices likely reaching new highs.

Qatar

Oil Exports: Qatar accounted for 2% of the world's oil exports last year.

Natural Gas Exports: Qatar is the fourth-largest exporter of natural gas, but the world's leading exporter of LNG. This gives Qatar greater flexibility in where it sends its natural gas than countries bound by pipelines. Qatar has two LNG complexes—Qatargas and Ras Laffan. In addition to its liquefaction capacity, Qatar also owns interests in LNG import terminals in Europe and the United States and one of the largest LNG tanker fleets. Qatar also exports natural gas by pipeline to the UAE.

Supplemental Energy Concerns: Qatar hosts of the Gas Exporting Country Forum's headquarters in Doha. It is also home to important U.S. military Central Command facilities that support ongoing operations in MENA, Iraq, and Afghanistan.

Saudi Arabia

Oil Exports: Saudi Arabia is the key oil-producing country in MENA, and by most industry measures the world. Besides being the largest exporter of oil, Saudi Arabia is the holder of most of the world's spare production capacity. Key within Saudi Arabia's petroleum complex are two pieces of infrastructure: Abqaiq Oil Processing Center and the Ras Tanura Oil Exporting Terminal. Both of these are essential parts of Saudi Arabia's ability to export oil. Oil is processed at Abqaiq in preparation for export or refining. Abqaiq has a 7 Mb/d oil processing capacity. Ras Tanura, the main export terminal for Saudi oil, has a 6 Mb/d export capacity. Should either of these facilities be shut down, the impact on oil markets would be significant, with prices likely rising to new highs very quickly. Abqaiq was targeted in a failed attack by terrorists in February 2006.

Natural Gas Exports: Although Saudi Arabia is one of the largest producers of natural gas, it does not export any natural gas. Natural gas is used domestically for industry and electric power generation.

Supplemental Energy Concerns: Saudi Arabia has played a key role in OPEC as the swing producer, increasing or decreasing its production to maintain prices. It has also been a moderate voice within the cartel.

Syria

Oil Exports: Syria is a minor exporter of oil.
Natural Gas Exports: Syria does not export natural gas.

Supplemental Energy Concerns: Syria is a transit country for the Arab Gas Pipeline from Egypt. Additionally, Syria could play a role in potential Iraq natural gas exports.

Tunisia

Oil Exports: Tunisia is a minor exporter of oil.

Natural Gas Exports: Tunisia does not export natural gas.

Supplemental Energy Concerns: Tunisia is a transit country for Algeria's Galsi natural gas pipeline to Italy.

United Arab Emirates (UAE)

Oil Exports: The UAE accounted for 5% of global oil traded in 2009 and had the second-largest spare capacity of OPEC members.

Natural Gas Exports: The UAE is a net importer of natural gas, receiving over 600 bcf of natural gas from Qatar via pipeline.

Supplemental Energy Concerns: The UAE is a global financial hub.

Yemen

Oil Exports: Yemen is a minor exporter of oil.

Natural Gas Exports: Yemen is a minor exporter of natural gas.

Supplemental Energy Concerns: Terrorist activity in Yemen has led to concern over its exports of LNG that was destined for the United States.

Yemen is a littoral country for a shipping chokepoint called the Bab el-Mandab, through which transited an estimated 3.2 Mb/d of oil last year cumulatively in both directions. The Bab el-Mandab is the southern entry/exit point to the Red Sea and therefore critical to much of the ship traffic that passes through the Suez Canal.

End Notes

[1] For detailed reports on political developments in the region, see reports at http://www.crs.gov/Pages/SubIssue.aspx? CLIID=282&ParentID=29.

[2] The price for West Texas Intermediate crude, the benchmark most often cited in the United States, has not increased by as much due to a variety of factors, including greater integration between the market for Brent and crudes from MENA. For details about recent developments in the Brent-WTI price spread, see http://www.eia.doe.gov/todayinenergy/detail.cfm?id=290.

[3] Christopher Johnson et al., "Libyan oil production, outage, exports, customers," *Reuters*, March 2, 2011.

[4] International Energy Agency, *Oil Market and Libyan Supply Update*, Paris, 2011, http://www.iea.org/files/ final_20110225_libyan_update.pdf.

[5] Sherry Su, "Saudi Arabia Said to Offer Europe Refiners Extra Oil This Month," *Reuters*, March 2, 2011.

[6] Gabriele Parussini, "IEA: Libya Unrest Starting to Hit Oil Supplies," *New York Times*, March 5, 2011.

[7] Jad Mouawad and Clifford Krauss, "Mideast Politics Again a Factor in Oil Prices; Despite Increased Output from the Saudis, Markets Fear the Spread of Unrest," *International Herald Tribune*, March 1, 2011.

[8] Spare capacity estimate is from the EIA's Short-Term Energy Outlook (*http://www.eia .gov/emeu/steo/pub/* contents.html). Note, only Members of the Organization of the Petroleum Exporting Countries (OPEC) hold spare production capacity. The cartel aims to adjust supply to manage oil prices, which requires holding some capacity in reserve. Saudi Arabia, Kuwait, and Libya are members. Egypt is not.

[9] Being a net importer of a particular good is not necessarily negative from an economic point of view. The United States imports many products because they would be more costly to consumers and businesses were they produced domestically. However, imports can become a drag on economic growth when the cost of imported goods rise, and greater import dependence can increase the degree to which a price shock disrupts the economy. Rising oil prices may have contributed to the U.S. recession in 2007-2008, according to James D. Hamilton, a Professor of Economics at the University of California, San Diego ("Causes and Consequences of the Oil Shock of 2007-08," Brookings Papers on Economic Activity, April 27, 2009, http://dss.ucsd.edu/~jhamilto/Hamilton_oil_shock_08.pdf).

[10] U.S. Congress, Senate Committee on Banking, Housing, and Urban Affairs, *The Semiannual Monetary Policy Report to the Congress*, 112th Cong., 1st sess., March 1, 2011.

[11] Dean Maki, Barclays Capital, as quoted by Motoko Rich et al., "Rising Oil Prices Pose New Threat to U.S. Economy," *The New York Times*, February 24, 2011.

[12] "Energy & Oil Prices," *Bloomberg*, February 28, 2011, http://www.bloomberg.com/energy/.

[13] U.S. Energy Information Administration databases (*http://www.eia.gov/dnav* /pet/pet_pri_spt_s1_a.htm).

[14] LNG is natural gas that is cooled to -260 °F, which decreases it in volumetric size to 1/600th, making it economical to transport by tanker.

[15] Natural gas tends to be priced in dollars per million British thermal unit ($/mmBtu) as in Figure 3. The Btu component of the units is to express the energy content as opposed to the volumetric units (billion cubic feet (bcf)) used in **Table 2**. To convert from 1 mmBtu to bcf multiply by 0.00099.

[16] OPEC members include Algeria, Angola, Ecuador, Iran, Iraq, Kuwait, Libya, Nigeria, Qatar, Saudi Arabia, the United Arab Emirates (UAE), and Venezuela. Iraq is not currently participating in OPEC's oil production quota agreements.

[17] GECF members include Algeria, Bolivia, Egypt, Equatorial Guinea, Iran, Libya, Nigeria, Qatar, Russia, Trinidad and Tobago, and Venezuela.

[18] Ulf Laessing, "Saudi King Back Home, Orders $37 Billion in Handouts," *Reuters*, February 23, 2011. Also "Oman Ups Minimum Monthly Wages to $520 for Nationals," *Reuters*, February 16, 2011.

[19] The SPR can hold a maximum of 727 million barrels of oil. Currently, the SPR contains 292.5 million barrels of sweet crude and 434 million barrels of sour. There have been calls for the SPR to hold refined products in addition to crude oil as some other members of the IEA have done. This was particularly raised after hurricanes Katrina and Rita shut down U.S. refineries in 2005.

[20] Like the United States, the other members of the IEA have agreed to hold strategic stocks and coordinate their use in the event of global supply disruptions. IEA members collectively hold 1.6 billion barrels in strategic stocks. Members are required to hold 90 days' worth of imports, a requirement met through a combination of government held and industry held stocks. The 726.5 million barrels mentioned above is held by the U.S. government.

[21] Refilling the SPR would also temporarily increase demand and put upward pressure on prices. The current average cost for oil in the U.S. SPR is $29.76 per barrel, well below current market prices and well below last year's average market price of $80 per barrel, according to DOE (http://fossil.energy.gov/programs/reserves/spr/spr-facts.html). It is unlikely that the new oil to refill the SPR would be that inexpensive.

[22] Office of Fossil Energy, U.S. Department of Energy, *Strategic Petroleum Reserve - Quick Facts and Frequently Asked Questions*, August 10, 2010, http://fossil.energy.gov /programs/reserves/spr/spr-facts.html.

[23] Crude has been sold from the SPR twice for major emergencies. In 1990/91, 21 million barrels were sold in response to supply disrupted by Iraq's invasion of Kuwait and U.S. Operation Desert Shield/Storm. By late 1990, the invasion had removed 4.6 Mb/d of oil from the market. The only other outright emergency sale was for 11 million barrels in 2005 after Hurricanes Katrina and Rita. More than 1 Mb/d of oil production was shut-in by the storms in September 2005. Oil has also been loaned from the SPR under exchange agreements on 10 occasions, many of them also involving smaller domestic supply disruptions. Apart from the crude sale after Hurricanes Katrina and Rita, 9.8 million barrels were also loaned out under exchange agreements. Separately, Congress also authorized the non-emergency sale of 28.1 million barrels from the SPR in 1996 to raise revenues. For more on the specific sales and exchanges of the SPR, see http://fossil.energy.gov/programs/reserves/spr/spr-drawdown.html#katrina_sale.

[24] Jennifer Dloughy, "Some Oil from Reserve to Be Sold," *Houston Chronicle*, February 16, 2011.

[25] U.S. Congress, Senate Committee on Energy and Natural Resources, *FY 2012 Budget for the Department of Energy*, To receive testimony on the U.S. Department of Energy's budget for fiscal year 2012 by Energy Secretary Chu, 112th Cong., 1st sess., February 16, 2011.

[26] Office of Senator Jack Reed, "Reed & Whitehouse Urge Obama to Swiftly Sell $500 Million of Emergency Oil from Faltering Cavern," press release, March 1, 2011, *http://reed .senate.gov/newsroom/details.cfm?id=331504*.

[27] Senator Jeff Bingaman, "World Oil Supplies," remarks in the Senate, *Congressional Record*, daily edition, vol. 157 (March 2, 2011), pp. S1106-1107.

[28] Letter from Representatives Edward J. Markey, Rosa L. DeLauro, and Peter Welch to The Honorable Barack Obama, President, February 24, 2011, *http://democrats* .naturalresources.house.gov/sites/ democrats.resourcescommittee.house.gov/files/2011-02-23_SPR_Letter_to_Pres_Obama.pdf.

[29] "US Says SPR Release 'Option We Are Considering,'" *Platts*, March 7, 2011.

[30] See CRS Report RL31720, *Energy Policy: Conceptual Framework and Continuing Issues*, by Robert Bamberger.

[31] "Idea of Gas-Tax Holiday Runs into a Dead End," *Associated Press*, July 19, 2008.

[32] For additional information on Egypt's impact on oil and natural gas markets, see CRS Report R41632, *Implications of Egypt's Turmoil on Global Oil and Natural Gas Supply*, by Michael Ratner.

[33] *BP Statistical Review of World Energy 2010*, BP, p. 28.

[34] For additional information on Israel's natural gas industry, see CRS Report R41618, *Israel's Offshore Natural Gas Discoveries Enhance Its Economic and Energy Outlook*, by Michael Ratner.

In: U.S. Oil Sources
Editor: Natalie B. Kelleher
ISBN: 978-1-62100-415-8

Chapter 5

IMPLICATIONS OF EGYPT'S TURMOIL ON GLOBAL OIL AND NATURAL GAS SUPPLY

Michael Ratner

SUMMARY

The change in Egypt's government will likely not have a significant direct impact on the global oil and natural gas markets. There may be some short-term movements in price, mostly caused by perceived instability in the marketplace, but these would most likely be temporary. However, prolonged instability that raises the specter of spreading to other oil and natural gas producers in the region would likely add to upward price pressures. Although Egypt is considered an energy producer or net exporter overall, its oil and natural gas exports are not large enough to affect regional or global prices. The most serious impact would be on regional recipients of its natural gas exports.

Egypt's main influence on energy markets is its control of the Suez Canal and the Suez-Mediterranean oil pipeline (SUMED). The current low utilization of these two pieces of infrastructure would likely limit any effect of their closure in the near term. Both the oil and natural gas industry would, over time, find alternative routes to circumvent the canal and pipeline if necessary.

INTRODUCTION

The potential outcome of the resignation of long-time Egyptian President Hosni Mubarak is unknown and any ramifications for the oil and natural gas sectors are uncertain.[1] This paper examines the impact of a disruption of Egypt's oil and natural gas sector or a complete halt to exports of either oil or natural gas and closure of the Suez Canal and the Suez-Mediterranean (SUMED) oil pipeline, and the impact of those actions on world oil and natural gas markets. It is important to keep in mind that even the most nationalistic, isolationist, or anti-Western government would most likely not undertake all these measures. Oil, natural gas, and transit generate large amounts of revenue for Egypt and taking these measures could precipitate outside intervention, particularly closing the Suez Canal.[2] Additionally, the timing of these actions would also change the impact on oil and natural gas markets, i.e., whether they occurred during the summer driving season or the winter heating season.

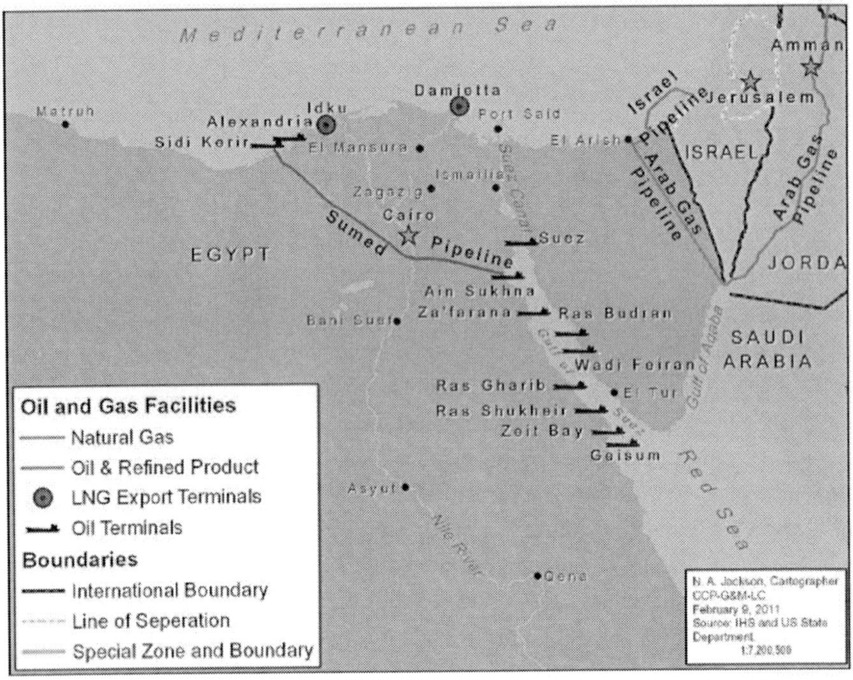

Source: CRS Cartography.

Figure 1. Egyptian Oil and Natural Gas Infrastructure.

2009 total: 559 million barrels of oil equivalent (mboe) produced and 695 mboe consumed

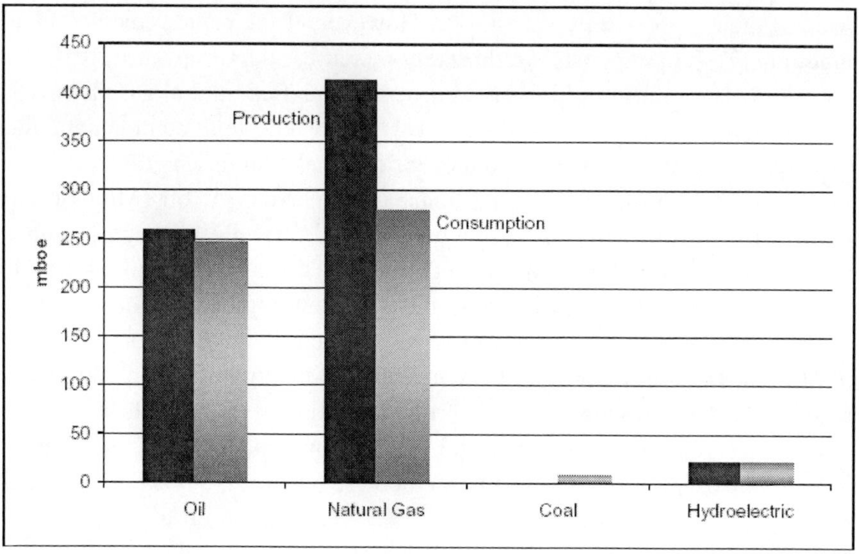

Source: BP Statistical Review of World Energy 2010, http://www.bp.com /sectiongenericarticle.do?categoryId=9033088&contentId=7060602.

Figure 2. Egyptian Primary Energy Production and Consumption.

An additional factor mitigating the impact on world oil and natural gas markets is that in 2009, Egypt consumed much of the energy it produced and had to import coal, highlighting the limited importance of Egypt as a global energy producer; see **Figure 2.**[3]

CLOSING OF THE SUEZ CANAL AND SUEZ-MEDITERRANEAN OIL PIPELINE

Closing the Suez Canal would be one of the most visible actions a new government could take, particularly for the oil and natural gas industry.[4] Although it would probably take only several weeks to re-route and re-size[5] oil and natural gas tankers along with a possible drawdown of inventories, a closure of the canal would cause an immediate and most likely short-lived rise in global oil prices. Despite there being adequate spare production capacity in the world, oil prices tend to react quickly to market disruptions before settling back to their pre-existing price range. However, if the canal remained closed indefinitely, shipping costs would likely increase, adding upward pressure on oil prices. The same would likely be true for liquefied natural gas (LNG), but the effect would be less. Most LNG is sold under long-term contract, and there is currently a glut of LNG around the world to make up for any disruption.

In 2009, an estimated 1.8 million barrels per day of oil (Mb/d) (of the world's roughly 80 Mb/d of production) moved through the canal—almost 1 Mb/d northbound and 0.85 Mb/d southbound.[6] This is down from 2.4 Mb/d in 2008. The decline is mostly attributed to lower global demand for oil; production cuts by the Organization of the Petroleum Exporting Countries (OPEC), particularly from the Persian Gulf producing countries; and piracy.[7] More oil is also flowing to Asia from the Middle East, while more West African oil is going to Europe and North America, and does not have to traverse the canal. 2010 data shows an increase in cargos, but is incomplete for the year. The last time the canal was closed, in 1967 until 1975,[8] approximately 60% of Europe's oil supplies had been passing through the canal. Currently, only about 15% is shipped through the canal.[9]

Similar to the decrease in cargos through the canal, the SUMED oil pipeline—which is owned through state companies by Egypt (50%), Saudi Arabia (15%), the United Arab Emirates (15%), Kuwait (15%), and Qatar (5%)—is operating below its capacity of 2.5 Mb/d. In 2009, approximately 1.1 Mb/d moved through the pipeline, a decrease of about 50% compared to 2008. The reduction resulted from many of the same causes as the canal's drop-off in oil transportation.

In 2009, 331 billion cubic feet (bcf) of LNG traversed the canal, which represents 4% of global LNG trade and less than 1% of natural gas consumed globally. Unlike oil, LNG cargos through the Suez Canal have been steadily rising. Between 2008 and 2009, the volume of LNG passing through the Suez Canal increased over 40%, more than doubling northbound cargos. Data for the first 10 months of 2010 show a 66% increase in LNG shipments through the canal compared to the 2009 total. The rise in LNG cargos is mostly attributable to the large increase in Qatar LNG exports. Nevertheless, there is ample supply of LNG in the global market, and rerouting LNG cargos to demand centers could be accomplished in a relatively short time frame.

Table 1. 2009 Egyptian Natural Gas Data
Units = billion cubic feet (bcf)

Region/Country	**Imports from Egypt (bcf)**	**% of Total Imports**	**% of Total Consumption**
Asia	36.73	<1%	<1%
Europe	235.20	1.5%	1.3%
• Spain	144.79	11.4%	12.0%
Middle East	194.23	17.6%	1.6%
• Israel[a]	60.03	100.0%	52.3%
• Jordan	100.65	100.0%	91.9%
• Lebanon	1.41	100.0%	100.0%
• Syria	32.14	100.0%	12.8%
North America	175.16	3.5%	<1%
• United States	160.33	4.3%	<1%
South and Central America	5.65	1.0%	<1%
• Argentina	5.65	6.0%	<1%
	Exports (bcf)	**Production (bcf)**	**Consumption (bcf)**
Egypt	646.97	2,213.88	1,566.91
% of World	2.1%	2.1%	1.5%

Source: BP Statistical Review of World Energy 2010, *http://www.bp.com/liveassets* /bp_internet/globalbp/ globalbp_uk_english/reports_and_publications/statistical_energy_review_2008/S TAGING/local_assets/ 2010_downloads/natural_gas_section_2010.pdf, and other industry sources.

a. For additional information on Israel's natural gas use, see CRS Report R41618, *Israel's Offshore Natural Gas Discoveries Enhance Its Economic and Energy Outlook*, by Michael Ratner.

NATURAL GAS SUPPLIES: A REGIONAL ISSUE

While Egypt is a relatively small player in the global natural gas industry, it can have a larger impact on regional natural gas supply. Egypt produces all the natural gas consumed in Lebanon, almost all the natural gas consumed in

Jordan, and more than half the natural gas consumed in Israel. On a world scale, Egypt accounted for only 2.1% of global natural gas production and 2.1% of global natural gas trade in 2009. Egypt holds 77 trillion cubic feet or 1.2% of the world's proved gas reserves.

Egypt exports natural gas through two pipelines and two LNG facilities. Currently, the Arab Gas Pipeline connects Egypt to Jordan, Lebanon, and Syria. There are plans to expand the pipeline further, enabling Egypt to export natural gas through Turkey to Europe.[10] In 2008, Egypt opened an export pipeline to Israel. There was an explosion in early February that shut down both pipelines for a short time although the damage was primarily to the Arab Gas Pipeline.[11] Egypt has almost 600 bcf of LNG export capacity, with one facility in Damietta and one in Idku. Egypt accounted for approximately 5% of global LNG trade last year, with most cargos going to Europe.

Recent Unrest

At the end of April 2011 the natural gas terminal near El-Arish in Egypt (see **Figure 1** above) was attacked for the second time since protests erupted in that country in January.[12] Natural gas from the terminal supplies the Arab Gas Pipeline to Jordan, Syria, and Lebanon, and a separate pipeline to Israel. There is no estimate for how long natural gas will not be exported. The pipeline was also attacked and disabled in February causing natural gas supplies to be stopped for about a month. The terminal has been a target for Bedouins who feel neglected and oppressed by Cairo.[13]

Oil Exports Already Minimal

Aside from its role as a transit center, withdrawal of Egypt's own oil production from the global market would likely have limited impact on world oil prices. In 2010, Egypt was a small net importer of oil, producing approximately 0.66 Mb/d while consuming close to 0.71 Mb/d.[14] Egypt's oil production has been in decline since the early 1990s, a trend not likely to be reversed.[15] The country—which has the largest refining capacity in Africa, with 975,000 barrels of processing capacity—did export some refined products, such as naptha, but imported others. Cutting off exports of naptha, an oil product that can be used in making petrochemicals and gasoline, could put minor upward pressure on European prices, which is a main market for Egyptian exports.

GLOBAL OIL AND NATURAL GAS MARKETS: A SHORT-TERM CONCERN

Global oil prices have already reacted to the unrest in Egypt despite no disruption to Egyptian production. The reaction comes from two concerns: (1) the near-term risk that any disruption to oil transit through Egypt could delay oil shipments for several weeks (as they are rerouted around the Cape of Good Hope, adding extra shipping time) and (2) the more significant concern that political unrest could spread to other, more important energy exporters in the region, such as Saudi Arabia, Iraq, or Kuwait. There has been some upward pressure on natural gas prices as companies scramble to hedge against a disruption of Egyptian exports, but the benchmark natural gas price in the United Kingdom is actually down.

Should the scenario described above come to pass, there would likely be little impact on the global oil and natural gas market in the long term. Both industries will take some time to recalibrate flows, but should be able to accomplish that with minor or no disruptions. Additionally, the strategic petroleum reserves of member countries of the International Energy Agency, which are mostly from the Organization for Economic Cooperation and Development (OECD), could also be utilized to bridge any gap in oil flows should a disruption prove to be significant.

End Notes

[1] For additional information on the current events in Egypt please refer to CRS Report RL33003, *Egypt in Transition*, by Jeremy M. Sharp.

[2] This paper does not discuss the internal impact of the scenario for Egypt, nor does it speculate on the global reactions of various governments.

[3] 2009 is the latest year for complete production and consumption data.

[4] In addition to oil and natural gas, many other commodities traverse the Suez Canal as well as U.S. and other nations' naval ships.

[5] Currently, Very Large Crude Carriers (VLCC) and Ultra Large Crude Carriers (ULCC) class tankers are unable to traverse the Suez Canal because of their size, and oil must be put on smaller ships or shipped through the SUMED pipeline between the Red Sea and Mediterranean. If the canal were closed, more oil would be transported by VLCCs and

ULCCs to make the journey around Africa's Cape of Good Hope, adding an additional 15 days of transit to Europe and 8-10 days to the United States Gulf Coast, according to the U.S. Energy Information Administration's country report on Egypt (see footnote 6 below).

[6] U.S. Energy Information Administration, *Country Analysis Briefs: Egypt*, February 2011. http://www.eia.doe.gov/ emeu/cabs/Egypt/pdf.pdf

[7] Marsoft, *Marsoft Flash Report: New Suez Canal Crisis?*, February 2011.

[8] The Suez Canal has only been closed twice in its 141-year history: for 17 months during the 1956 Suez Crisis, and for almost eight year after the 1967 Six Day War with Israel.

[9] Marsoft, February 2011.

[10] Egyptian natural gas would feed into the planned Nabucco pipeline or other natural gas pipelines. See CRS Report RL34642, *Turkey: Selected Foreign Policy Issues and U.S. Views*, by Jim Zanotti, for more on the Nabucco pipeline project proposal.

[11] Batsheva Sobelman, "Israel: Egypt Gas Pipeline Explosion Raises Energy Concerns," *Los Angeles Times*, February 5, 2011.

[12] Ashraf Sweilam, Explosion rocks Egypt gas terminal near Israel, *Associated Press*, April 27, 2011.

[13] Ibid.

[14] U.S. Energy Information Administration, *Country Analysis Briefs: Egypt*, February 2011.

[15] Ibid.

In: U.S. Oil Sources
Editor: Natalie B. Kelleher
ISBN: 978-1-62100-415-8

Chapter 6

HEDGE FUND SPECULATION AND OIL PRICES

Mark Jickling and D. Andrew Austin

SUMMARY

Dramatic swings in crude oil prices have led Congress to examine the functioning of the markets where prices are set. A particular concern is that financial speculators may at times drive prices above the level justified by supply and demand. Most oil speculators do not produce or take delivery of commercial quantities of oil; rather, they trade financial contracts whose value is linked to the price of oil. These derivative contracts—futures, options, and swaps—allow speculators to profit if they can forecast price trends or exploit new arbitrage opportunities. Derivatives also permit oil companies, airlines, utilities, and other energy-consuming or energy- producing firms to reduce or "hedge" price risk by locking in today's price for transactions that will occur in the future. Hedgers and speculators trade on regulated futures exchanges in a continuous auction market. Prices set there serve as benchmarks for many physical oil transactions.

Some contend that oil speculators do not always trade on fundamental information related to supply and demand, but are nonetheless able to drive up prices by flooding the market with cash and overwhelming the influence of commercial hedgers who actually deal in physical oil. On the other hand, most empirical studies suggest that speculation does not generally increase price volatility, although the occasional emergence of speculative "bubbles," when market prices may diverge significantly from fundamental values, is well known. Neither economists nor regulators have reached a consensus as to the causes of oil price movements in recent years—some point to the fundamentals (where both demand and supply are inelastic in the short run, and questions exist about the capacity to meet growing global demand in the long run), while others focus on financial markets. Both are possible sources of price volatility.

This report examines the relationship between the price of oil and the positions of various classes of traders in crude oil futures and options. Position data come from the *Commitments of Traders* report, published weekly by the Commodity Futures Trading Commission (CFTC). A statistically significant correlation is evident between changes in positions held by "money managers" (a category of speculators that includes hedge funds) and the price of oil. In other words, during weeks when money managers have been net buyers of oil futures and options (or increased the size of their long positions), the price has tended to rise. Price falls, conversely, have tended to coincide with reductions in money managers' long positions. This statistical relationship is weaker for other classes of speculators and for commercial hedgers.

There are several possible explanations why money managers' trades might be more closely linked to prices. Money managers might identify information that will affect prices (and trade on that information) more quickly or accurately than other market participants. Other traders might copy the trades of certain hedge funds viewed as market leaders, driving prices further in the same direction. In this way, money managers' trades could move the price, even though their positions account for a relatively small share of the total market.

Causation could also run in the opposite direction—perhaps on average money managers chase price trends rather than set them. Other traders whose positions appear in the *Commitments of Traders* data, such as commercial hedgers or swap dealers, may be less price-sensitive than hedge funds and react more slowly to price changes. Data presented in this report cannot explain causes of oil price movements, but are intended to provide a context for evaluating arguments about the impact of speculation.

OIL PRICE SPIKES AND SPECULATION

The price of crude oil rose sharply in the first half of 2008, sparking widespread concern about energy prices, their effect on the world economy, and consequences for household consumption. Prices then fell sharply between July and December, from a peak of $145 per barrel to just over $30. To some observers, fundamentals of the oil market and macroeconomic conditions provide an adequate explanation for these price movements—the financial crisis deepened in the summer of 2008, leading to sharply reduced estimates of economic growth in much of the world and lowered projections of energy demand. Others, however, doubt that supply and demand conditions justified an 80% price plunge and argue that financial speculators had created artificially high prices.

In the first few months of 2011, crude oil prices again rose sharply, although they have fallen in May and June. These price movements have revived the debate on whether oil price trends are mostly driven by speculation or by changes in supply and demand fundamentals.

Economists and market regulators have not reached a consensus as to the causes of oil price movements in recent years. A number of studies attribute volatility to such supply and demand factors as turmoil in oil-producing countries, reduced production in some major oil fields, and the growth of demand from China, India, and industrializing middle-income countries.[1] An interagency task force led by the Commodity Futures Trading Commission (CFTC), found that "the increase in oil prices between January 2003 and June 2008 [was] largely due to fundamental supply and demand factors" and that "analysis to date does not support the proposition that speculative activity has systematically driven changes in oil prices."[2]

Others, such as CFTC commissioner Bart Chilton, have contended that oil price gyrations are likely the result of speculative trading.[3] A frequent argument has been that increasing investment flows from financial investors have affected prices.[4] Some analysts have sharply criticized those claims.[5] One econometric analysis that incorporated oil supply and energy demand effects concluded that speculation did not explain increases in oil prices in the 2003-2008 period, although the study suggested that speculation may have played some role in previous oil price spikes.[6]

The role of speculators in oil and other commodity markets has also attracted congressional interest. For example, the staff of the Permanent Subcommittee on Investigations of the Senate Committee on Homeland Security and Government Affairs has found that excessive speculation has had "undue" influence on wheat price movements[7] and in the natural gas market.[8] A 2011 report by the minority staff of the House Committee on Oversight and Government Reform argues that "addressing excessive speculation offers the single most significant opportunity to reduce the price of gas for American consumers."[9]

On June 20, 2011, Senator Maria Cantwell announced that the Federal Trade Commission (FTC) had notified her that it had opened an investigation of anticompetitive practices in crude oil and petroleum product markets.[10] The FTC has conducted several prior analyses of oil, gas, and petroleum markets.[11] The Energy Independence and Security Act of 2007 (EISA; P.L. 110-140) gave FTC broader powers to prevent market manipulation in the wholesale petroleum industry. In August 2009, FTC issued a Final Rule on Prohibitions on Market Manipulation, which became effective on November 4, 2009.[12]

Legislation before the 112th Congress (S. 1200) would authorize and direct the CFTC to take certain actions to reduce the volume of speculation in oil and related energy commodities.

This report provides background on futures and options markets for crude oil and presents data analysis of a possible relationship between market activities of speculative traders and oil prices.

The data presented in this report cannot explain causes of oil price movements, but are intended to provide a context for evaluating arguments about the impact of speculation.

Crude Oil Contracts

A crude oil futures contract is an agreement to buy or sell 1,000 barrels of oil at some future date at a price set today. Thus, the contract gains or loses value as prices fluctuate. A contract to buy oil (called a long contract) gains value if the price rises, because the holder is entitled to buy at the old, lower price. Conversely, a short contract requires the holder to sell at today's price, and gains value if prices fall, because the holder may sell at above the market price. A long position in futures may be described as a bet that prices will rise; a short position is a bet that they will fall. Each futures contract has a long and a short side—whatever one trader gains, the other loses.

Hedgers use futures not to bet on the price, but to avoid price risk. For example, a long contract in effect provides insurance to an oil refinery against an increase in the price of crude oil; if prices rise, the firm will pay more for oil on the physical (or "spot") market, but appreciation in the futures position offsets the price increase. Thus, the firm can use futures to lock in the price that prevailed when it entered into its position.[13]

Futures contracts and options on futures are traded on a number of exchanges around the world, linked to several different grades of crude oil.[14] The most popular crude oil contract is traded on the New York Mercantile Exchange (or Nymex, now part of the CME Group, Inc.). The contract represents 1,000 barrels of West Texas Intermediate (WTI) grade crude oil, deliverable at Cushing, Oklahoma. Although Nymex WTI contracts call for physical delivery of 1,000 barrels at the time the contract expires, in practice nearly all contracts are settled for cash, without either party taking or making delivery.[15] A trader may exit the market at any time by simply purchasing an offsetting position. That is, the holder of a long contract purchases a short contract with the same expiration date—since his obligation is then to buy and sell the same commodity at the same time, his net exposure is zero, and he is said to be "evened up," or out of the market.

Nymex offers an oil contract expiring each month through the end of 2016. Most trading is in the contract expiring soonest, called the "near month." An identical WTI crude contract trades on the ICE Futures Europe exchange in London.

Most trading on the futures exchanges is short-term. Many traders prefer to even out their positions before the close of trading each day in order to avoid overnight price risk. Some traders, however, take longer-term positions, either to hedge transactions expected to take place months or years in the future, or to speculate on long-term price movements. The number of contracts outstanding at the end of the trading session is called the "open interest." The Commitments of Traders (COT) report published by the Commodity Futures Trading Commission (CFTC) shows the open interest in futures and options on futures, broken down by several classes of market participant, distinguishing between commercial hedgers and speculators. These data, though limited in important ways (discussed below), are the best public source of information on the activity of speculators in the crude oil market.

COMMITMENTS OF TRADERS (COT) REPORTS

COT data, usually published each Friday in the late afternoon, reflect the open interest, or the number of contracts outstanding, as of close of trading on the previous Tuesday. Thus, comparing week-to-week COT figures shows whether classes of traders have increased or decreased the size of their long, short, or spread positions.[16] The COT figures do not show how much trading has gone on during the week, or whether a position has been liquidated and then built back up, but simply offer a snapshot of positions at the market close on Tuesday.

Another significant limitation of COT data is that they do not cover swap contracts—another form of oil derivative contract not traded on exchanges. Thus, COT figures arguably cover only a subset of oil derivatives, all of which play a role in setting prices. The Dodd-Frank Wall Street Reform and Consumer Protection Act (P.L. 111-203) gave the CFTC new regulatory authority over the swaps market. In the future, COT reports may reflect swap positions, but the data surveyed in this report cover only exchange-traded futures and options on futures.

COT data show the aggregate positions of hedgers and several classes of speculators, as follows:

- **Producer/Merchant/Processor/User.** These are entities that predominantly engage in the production, processing, packing, or handling of crude oil and use the futures markets to manage or hedge risks associated with those activities.
- **Swap Dealers** are entities that deal primarily in swaps for a commodity and use the futures markets to manage or hedge the risk associated with those swap transactions.[17] The swap dealer's counterparties may be speculative traders, like hedge funds, or traditional commercial clients that are hedging risk arising from their dealings in the physical commodity. Thus, swap dealer positions represent both hedging and speculation.[18]
- **Money Managers,** for the purposes of the COT reports, are registered commodity trading advisors (CTAs), registered commodity pool operators (CPOs), or unregistered funds identified by the CFTC. These traders, which include hedge funds, manage and conduct futures trading on behalf of clients.
- **Other Reportables.** Every other reportable trader[19] not placed into one of three categories above is put into the "other reportables" category. These may include floor traders and exchange members who trade for their own accounts.
- **Nonreportable** positions include all traders whose holdings are below the 350- contract reporting threshold. These may be hedgers or speculators.

Table 1. Crude Oil Open Interest: May 24, 2011
Futures and Options on Nymex and ICE Futures Europe

Type of Trader	Number of Contracts (1,00 0s)	Percent of Total	Number of Traders with Reportable Positions	
			Nymex	ICE
Producer/Merchant— Long	490	10.9	51	36
Producer/Merchant— Short	730	16.2	60	36
Swap Dealers—Long	275	6.1	17	15
Swap Dealers—Short	371	8.2	30	12
Swap Dealers—Spread	871	19.3	42	25
Managed Money—Long	298	6.6	87	21
Managed Money—Short	33	0.7	47	4
Managed Money—Spread	379	8.4	94	16
Other—Long	115	2.5	88	14
Other—Short	77	1.7	50	16
Other—Spread	660	14.6	120	30
NonReportable—Long	124	2.7	NA	NA
NonReportable—Short	91	2.0	NA	NA

Source: CFTC, Commitments of Traders report.

Table 1 above shows the breakdown of open interest in crude oil futures and options, for Nymex and ICE Futures Europe WTI crude contracts, as of May 24, 2011. The data in Table 1 suggest that speculators account for most of the open interest in crude oil contracts. The long and short hedging positions of the "Producer/Merchant" category accounted for only 27.1% of open interest.

Some observers believe that the limited presence of commercial hedgers itself indicates excessive speculation in the market.[20] The proportions by themselves, however, do not appear to explain recent oil price changes because they have been relatively constant since mid-2006, when the disaggregated COT data series begins.[21] Indeed, the total level of open interest has been relatively constant since 2006, as the data in **Figure 1** and **Figure 2** indicate. Thus, the data do not appear to support an argument that recent oil price spikes are the result of large, sudden inflows of speculative funds.

Figure 1 and Figure 2 plot total crude oil open interest against the price of oil between August 2007 (when oil prices began to rise to their 2008 peak) and May 2011. **Figure 1** ends on July 28, 2009; **Figure 2** begins the following week. **Figure 1** reports only Nymex positions, whereas **Figure 2** shows the sum of Nymex and ICE Futures Europe positions (which the CFTC did not report earlier). As noted above, the ICE and Nymex contracts are identical, so positions on the two exchanges have been added together.

While the total level of open interest appears to be more stable than the price of oil, the aggregate figures mask significant fluctuations in the size of positions held by different classes of traders. **Figure 3 and Figure 4** break down open interest positions for the various types of traders holding reportable positions. The time periods covered are the same as in **Figure 1 and Figure 2Figure 2. Figure 3** shows Nymex positions; **Figure 4** shows Nymex plus ICE.

Even though the aggregate open interest fluctuates within fairly narrow bounds, the number of contracts held by particular classes of traders may change significantly, as **Figure 3 and Figure 4** indicate. For example, commercial hedgers' positions (both long and short) shrank significantly beginning in June 2008, when oil prices neared their peak. Between June 10 and August 26, 2008, combined long and short positions of commercial hedgers fell by 49.7%. Meanwhile, the price of oil peaked at $145.32 per barrel on July 3, 2008, stood at $116.31 on August 26 (when the size of hedgers' positions reached a low point), and ultimately bottomed out at $30.28 on December 23, 2008.

August 2007 through July 2009

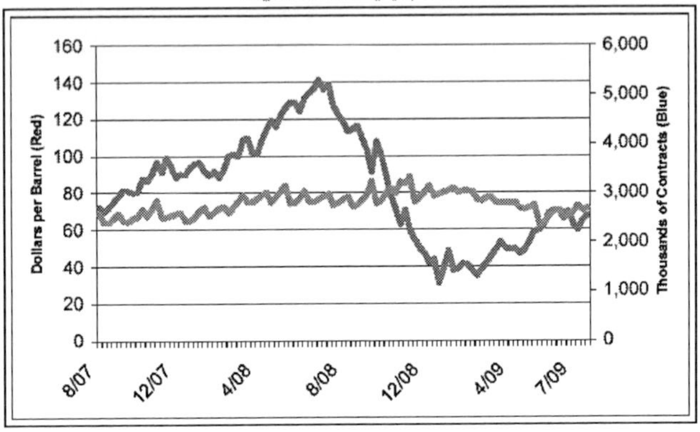

Source: CFTC, Commitments of Traders, and Global Financial Data.

Figure 1. The Price of Oil (Left Axis) and Nymex Open Interest (Right Axis).

In another example, money managers' long positions averaged 211,000 contracts during the second quarter of 2010, but increased by 46.4% to an average of 309,000 during the first quarter of 2011, when there was a significant upswing in the price of oil.

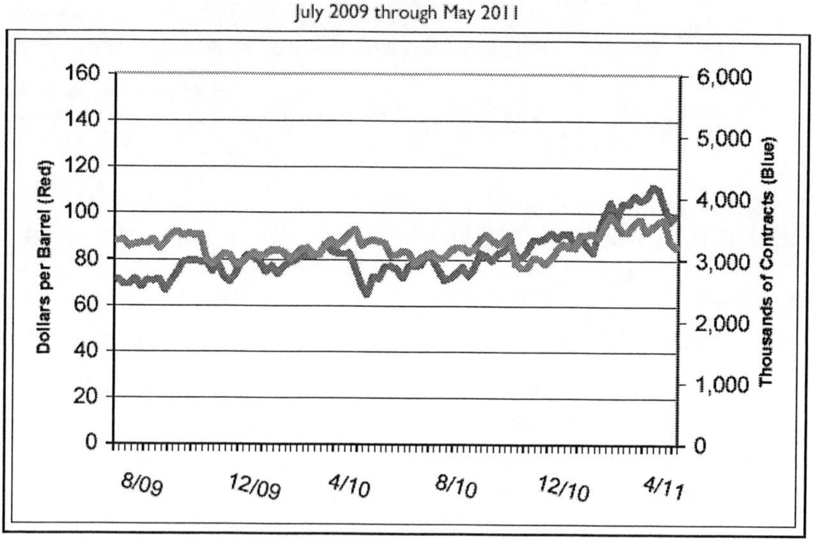

Source: CFTC, Commitments of Traders, and Global Financial Data.

Figure 2. The Price of Oil (Left Axis) and Open Interest on Nymex and ICE Futures Europe (Right Axis).

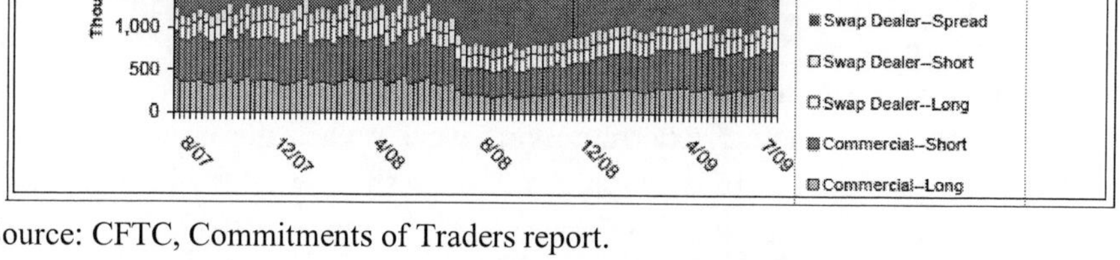

Source: CFTC, Commitments of Traders report.

Figure 3. Nymex Crude Oil Futures and Options in Open Interest, by Type of Trader.

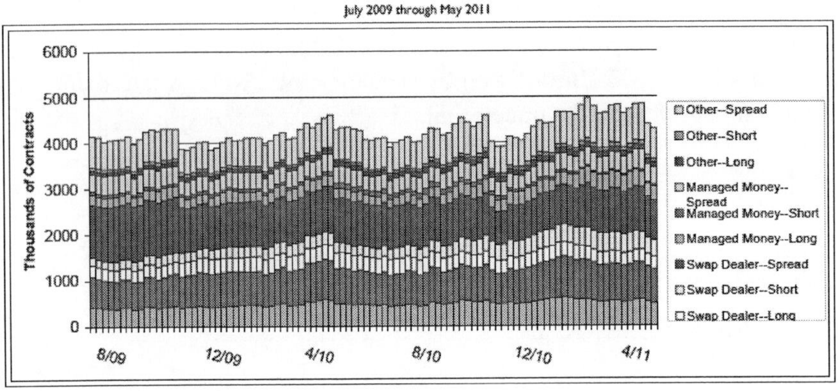

Source: CFTC, Commitments of Traders report.

Figure 4. Nymex and ICE Futures Europe Crude Oil Futures and Options in Open Interest, by Type of Trader.

Are changes in futures positions statistically related to price changes? Is the relationship the same for different classes of traders? The next section of this report examines how changes in crude oil futures and options positions correlate with changes in price.

CHANGES IN FUTURES MARKET POSITIONS AND PRICE TRENDS

Scatter Plots of Changes in Positions and Oil Prices

Analysis of COT data suggests that week-to-week changes in managed money long positions are positively correlated with changes in oil prices. **Figure 5** shows a scatter plot between week-toweek changes in managed money long positions and weekly changes in oil prices.[22] Each point in the figure represents a single week's data: changes over the previous week in the size of the long position are shown on the X-axis, price changes on the Y-axis. Points below zero on the Y-axis represents weeks when the price has fallen, while points to the right of zero on the X-axis show weeks when positions have increased. Thus, points in the lower left quadrant of the graph represent weeks when the price fell and money managers reduced their short positions. Points in the upper right indicate that prices rose that week and money managers increased their long positions.

The distribution of the points in **Figure 5** suggest a trendline rising from left to right. In other words, price rises are positively correlated with increased long positions, price falls with smaller positions.

Figure 6 shows a similar scatter plot for managed money short positions. Week-to-week changes in managed money *short* positions show a weaker negative correlation with changes in oil prices. In other words, money managers tended to increase short positions, which gain value as prices fall, when prices did fall, although the trend is not as clear as on the long side shown in **Figure 5**.

Figure 7 shows a scatter plot with changes in swap dealers' long positions, and **Figure 8** shows a scatter plot with changes in producer/merchant long positions. Note that while the changes in producer/merchant long positions are generally much larger than for managed money long positions, changes in producer/merchant long positions are not correlated with oil price changes.[23]

Scatter plots make visible the relationship between two variables and are especially useful in identifying outlier observations, which can have disproportionate influence on statistical models. For example, outliers seen in **Figure 5 and Figure 6** include the week following the Lehman Brothers bankruptcy in September 2008 and the weeks just after oil prices reached their lowest level in years in late December 2008.[24]

Weekly data, June 2006-May 2011

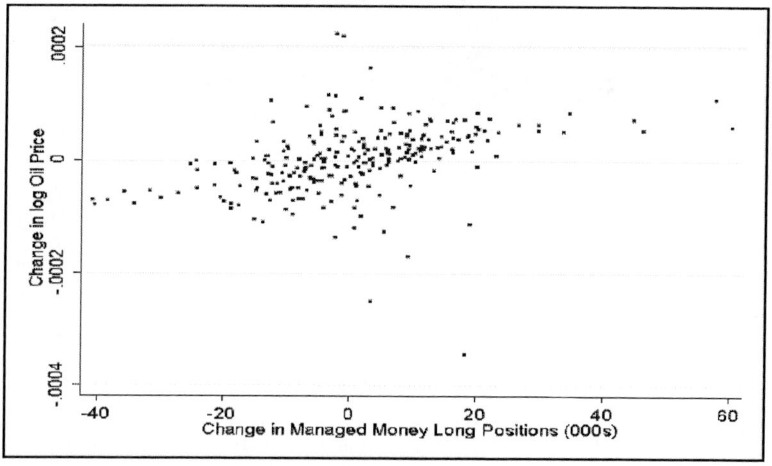

Source: CFTC and Global Financial Data (see text).
Note: Each dot represents data for one week.

Figure 5. Changes in Managed Money Long Positions and (log) Oil Prices.

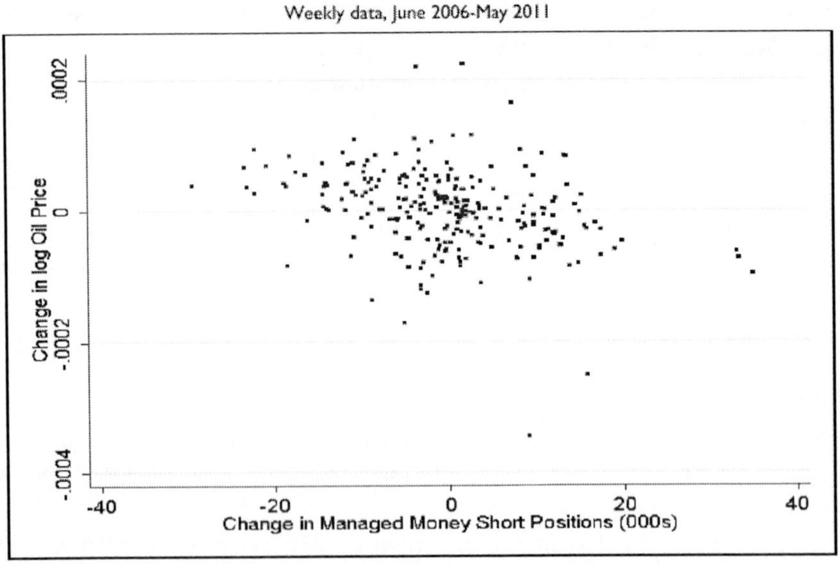

Source: CFTC and Global Financial Data (see text). Note: Each dot represents data for one week.

Figure 6. Changes in Managed Money Short Positions and (log) Oil Prices.

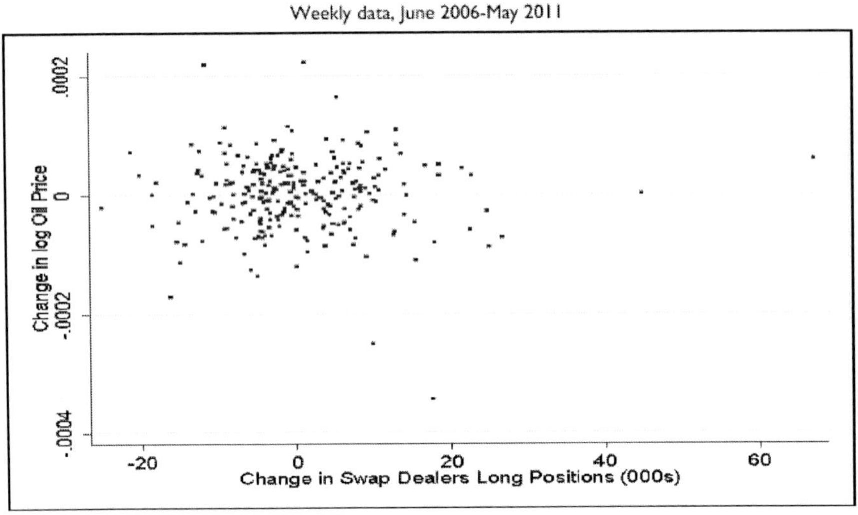

Source: CFTC and Global Financial Data (see text). Note: Each dot represents data for one week.

Figure 7. Changes in Swap Dealers' Long Positions and (log) Oil Prices.

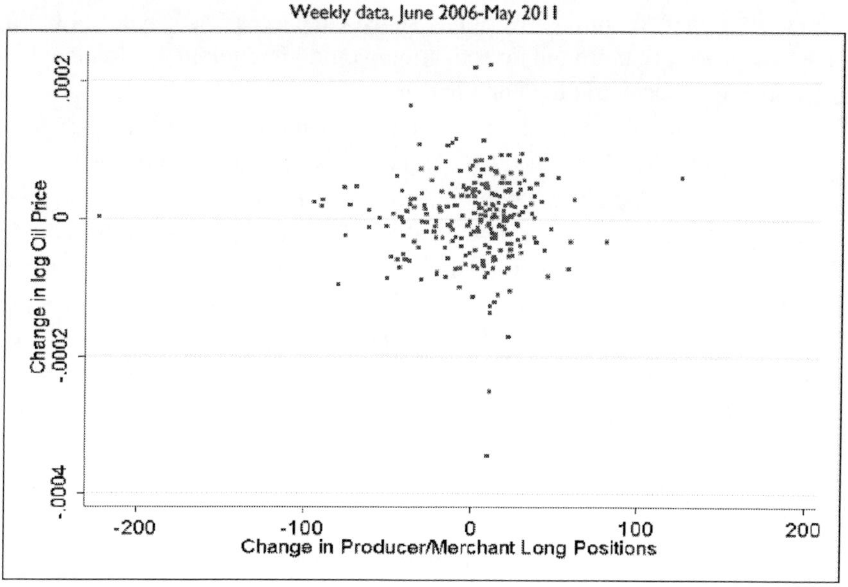

Source: CFTC and Global Financial Data (see text). Note: Each dot represents data for one week.

Figure 8. Changes in Producer/Merchant Long Positions and (log) Oil Prices.

One limitation of scatter plots is that the effects of control variables are not shown. An evident correlation between two variables could result from the influence of a unseen third, confounding variable. Alternatively, a relationship between two variables might only emerge after appropriate control variables were added. Correlations evident in **Figures 5** and **6**, however, also appear in regressions that account for the time-series nature of the data and control for seasonal variations, interest rates, and other financial variables.[25]

Regression Results

This section presents results from a time-series regression, reported in the **Appendix**, used to investigate possible effects of trading positions on the price of oil. This analysis uses the same weekly COT data discussed in the report, along with monthly seasonal controls and other financial controls.[26] Correlations between changes in managed money positions and percentage changes in oil prices are robust, and appear in a variety of specifications.[27]

While this analysis addresses some statistical issues associated with time-series data, a more thorough investigation would be required to disentangle the relationship between trading positions, prices, and macroeconomic conditions.

Control variables include a measure of bond yields and the euro/dollar exchange rate. The Corporate AAA Bond index may signal changes in market expectations of long-term economic growth. Oil investments may have been used as hedges against exchange rate risks for the U.S. dollar. In addition, fluctuations in the euro/dollar exchange rate affect the relative price of oil in Europe. Percentage changes in the euro/U.S. dollar exchange rate are therefore included as a control. Control variables are measured as changes or as percentage changes.

Results show a statistically significant positive correlation between managed money long positions and oil prices. The negative correlation between managed money short positions and oil prices is about as large, although less precisely estimated.

Swap dealers' positions show a similar pattern, although the estimated size of the effects are about two-thirds smaller and are less precisely estimated.

While changes in long and short positions of swap dealers and producer/merchant hedgers are not correlated with price changes when considered individually, controlling for changes in short positions does produce positive correlations between long positions and prices for swap dealers and producer/merchants, as well as for money managers. When controlling for long position changes, short position changes appear to be correlated to prices, and vice versa. These statistical relationships are discussed further in the **Appendix**.

The correlations between managed money long and short positions and oil prices, as estimated with these controls, are about twice as strong as those for producer/merchant traders. That is, an increase in managed money traders' long or short positions is associated with a percentage price change twice as large as for a change in producer/merchant positions. For swap dealers' position changes, the effect on prices is about 30% larger than for producer/merchant trades. While all of the position changes are estimated to have statistically significant effects, those for managed money positions, which are those held by professionally-managed investment vehicles such as commodity pools and hedge funds, are more precisely measured.[28]

Coefficients for corporate bond yields and the euro/dollar exchange rate are also statistically significant and have the expected signs.

INTERPRETING THE RESULTS

Correlation between changes in managed money positions and oil price movements, even with the introduction of control variables, does not show causation. Prices might rise because money managers buy, or trading decisions by money managers may reflect price changes. To use the terminology of a classic study on hedging and speculation, are money managers driving the price of oil, or are they hitchhikers along for the ride?[29] Interpretations on either side of the question are plausible.

Do Hedge Funds Cause Oil Price Changes?

The argument that money managers cause price movements faces certain difficulties. Any increase in a long position is equivalent to a purchase of the underlying commodity and thus tends to raise the price. However, money market long positions are a fairly small fraction of the total open interest positions.[30] That adjustments to this small market share would trigger discernable price movements might be considered surprising, especially when comparable or larger position changes by other market participants are not correlated with price swings.

One hypothesis is that money managers' trades may have a unique impact on intraday trading, which the weekly open interest data fail to capture. Short-term traders might observe and seek to copy the strategies of certain money managers who are regarded as especially capable of identifying new information that might be expected to move prices, or who simply have achieved superior returns in the past.[31] If significant numbers of short-term speculators copy money manager trades, the impact of those trades on prices would be magnified. In effect, under this scenario, money managers may have market power beyond what the size of their positions would suggest.[32]

Under this interpretation of the data, it would not matter whether hedge funds trade based on relevant fundamental information or not. If their trades trigger a significant number of similar transactions by others, they become a kind of self-fulfilling prophecy. Such "herding behavior" among speculators, if it exists, would support arguments that the oil price often includes a "speculative premium" above and beyond the price justified by the fundamentals.

An alternative explanation is that money managers do trade on fundamental information and that they are more adept than others at identifying information that is going to move prices. If money managers are consistent in their ability to identify new and relevant information that will affect prices (and trade on that information before others do), one result would be the observed correlation. A potential objection to this explanation is that it implies that some financial speculators are better analysts of the oil market than actual producers and end-users of oil, who also trade in the futures market.

Do Price Changes Cause Hedge Fund Trades?

Money managers might also profit by following price trends. Rather than cause price changes, they may buy when prices are rising and sell when they fall. But why would money managers' trading patterns, and not those of other market participants, be correlated with price changes in this way?

Other market participants may have longer investment time horizons or be less sensitive to price changes. Hedgers, for example, are generally less affected by price changes, because whatever they may lose on their futures positions, they make back in the spot market (because, for example, the physical commodity they produce will have gone up in price). Similarly, swap dealer positions may reflect long term index investments by pension funds and other institutional investors who are seeking to allocate part of their portfolio to an asset class that is not correlated to other assets they hold, such as stocks and bonds.[33] Since the object of such investment is portfolio diversification, such investors are less likely to buy or sell in reaction to short-term price movements.

Hedge funds, by contrast, are known for taking aggressive positions in search of high yields and for seeking to extract the maximum return from any price trend. A 2008 CFTC study referred to speculators "who take positions based on price expectations over a period of days, weeks, or months" as "trend followers."[34] Trading with this time horizon would be captured by the weekly COT reports, and may be more typical of money managers than other traders in oil futures and options.

Another possibility is that swap dealer COT data may represent the net of proprietary and customer positions, potentially masking the relationship between positions and price. It may be that swap dealers' trading for their own accounts would be correlated with prices, just as managed money positions are, but their trading on behalf of customers may follow a different pattern.

Money managers may follow prices in other ways. Some hedge funds apparently use "momentum trading" strategies to identify pricing blips that signal market activities of traders who are informed, but lack the backing of major financial resources.[35] For example, a scientist may understand the implications of a new technology, but may lack the ability to capture the full value of that information in markets. If that scientist bought shares of a company based on that knowledge, a hedge fund might infer something about that scientist's realization. The hedge fund, however, would be able to harness substantial financial leverage to exploit that information.

Certain hedge funds trading in oil markets might thus bootstrap information gleaned from trades of specialist traders who lack the financial resources to take full advantage of their informational advantages. Some contend that hedge funds and other sophisticated traders have used advantages in trade execution to benefit from slow-footed rivals.

DOES HEDGE FUND SPECULATION RAISE POLICY ISSUES?

The potential of money managers' trades to move prices does not necessarily raise any policy issue. If some commodity trading advisors or hedge funds are better or faster than others at trading on new information, the effect should be to make pricing more efficient. If too much "copycat" trading occurs, the market price may overshoot, but it is not clear why any such overshooting would not soon correct itself. One study found that open interest positions are highly correlated with macroeconomic activity.[36] This might suggest that traders with better information on macroeconomic trends, which strongly influence energy demand, take more aggressive positions, which would then influence oil prices. This interpretation suggests that speculative traders may have better information about market fundamentals, such as expectations of future economic growth, and thus help channel the effects of changes in current and future energy demand into market prices.

Some argue that money managers who are not involved in the physical oil business trade on information that is not fundamental and that this distorts prices. This effect would be exacerbated by copycat trading. Specifying which information is fundamental and which is irrelevant noise, however, is difficult. Experts may express opinions about what the fundamental price should be, given current supply and demand conditions, but a basic axiom of classical economics is that free markets do a better job of weighing information and determining prices than any group of experts.

On the other hand, asset markets can be susceptible to price bubbles, in which rising prices convince some traders that future prospects are rosier than previously thought.[37] Some have contended that identifying asset bubbles before a readjustment to fundamental levels (i.e., a crash) may be difficult.[38] Even if some traders understood that asset values exceeded fundamental values, it might be impossible given financing constraints to exploit that information to a degree sufficient to prevent the formation of bubbles.

Policy tools to address speculation that is viewed as excessive or destabilizing are generally blunt instruments. It is possible to prohibit trading in derivative instruments based on oil (or other energy commodities). With the Onion Futures Act (P.L. 85-39), Congress enacted a ban on onion futures, which remains in force, although it is not clear that onion prices have become more stable as a result.[39]

Other policy options (included in S. 1200) are steps to reduce the size of speculative positions, either by raising the costs of speculative trading or by limiting the number of contracts that speculators can hold. These measures could certainly reduce the volume of speculative trading in oil, but it is not clear that the effect would be to lower prices, for two reasons. First, the curbs would fall on short as well as long traders, making the net effect unpredictable. Second, oil is a global commodity, and a parallel futures market exists in London. ICE Futures Europe trades contracts based on U.S. oil, suggesting that oil speculation would continue despite any U.S. restrictions.

APPENDIX. STATISTICAL RELATIONSHIPS BETWEEN CHANGES IN POSITIONS AND PRICE

The regression results presented below show the correlation between changes in long and short positions held by money managers, swaps dealers, and commercial hedgers (producer/merchants).

While producer/merchant long and short positions are uncorrelated with price changes when considered individually, they are jointly correlated with price changes. Moreover, producer/merchant traders' long and short positions strongly correlate with each other, as **Figure A-1** shows. This result is counterintuitive, since long and short hedgers face opposite price risks. That is, when prices are rising, or expected to rise, long hedgers such as airlines or utilities would appear to have greater incentives to hedge the risk of further price rises, while short hedgers, such as oil companies, might be expected to reduce their hedging position to reap the full benefits of higher prices. But in fact, long and short hedgers appear to change their positions in tandem, whatever the prevailing price trend. (An example is discussed in the text: when oil prices peaked and began to fall in mid-2008, both long and short hedgers reduced the size of their positions.)

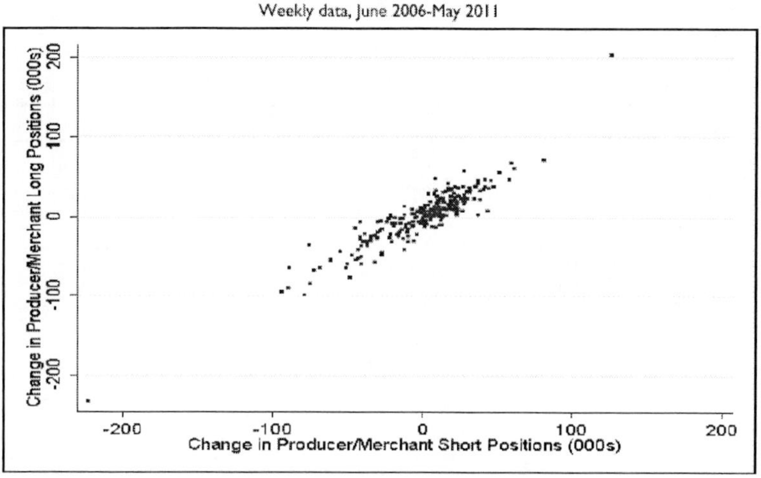

Source: CFTC and Global Financial Data (see text).

Figure A-1. Producer/Merchant Long and Short Positions.

Table A-1. Autoregression (AR) Changes in log Oil Prices Weekly data, June 20, 2006 through May 24, 2011

	Coefficient Estimate*	Standard Error*	z statistic	p-value	95% Confidence Interval	
					Low*	High*
Producer/Merchant Positions (Change, 000s)						
Long	0.953	0.313	3.05	0.002	0.34	1.57
Short	-0.999	0.308	-3.25	0.001	-1.60	-0.40
Swap Dealer Positions (Change, 000s)						
Long	1.290	0.386	3.33	0.001	0.53	2.04

Table A-1. (Continued)

	Coefficient Estimate*	Standard Error*	z statistic	p-value	95% Confidence Interval	
					Low*	High*
Short	-1.330	0.423	-3.15	0.002	-2.16	-0.50
Managed Money Positions (Change, 000s)						
Long	2.14	0.257	3.33	0.001	1.64	2.65
Short	-2.05	0.402	-3.15	0.002	-2.83	-1.26
A Corporate Bond Index (Change in logs)						
	0.710	0.335	2.12	0.034	0.05	1.37
Euro/U.S. $ Exchange Rate (Change in logs)						
	1.084	0.386	2.81	0.005	0.33	1.84
Lag Terms						
one-week lag	-0.180	0.155	-1.16	0.247	-0.48	0.12
two-week lag	-0.261	0.142	-1.84	0.065	-0.54	0.02
Wald chi2(21) = 430.38						
Prob > chi2 = 0.0000						
Number of obs =						

Source: Calculated by CRS using data from the CFTC and Global Financial Data.

Notes: * Magnitudes for positions variables coefficients, standard errors, and confidence interval bounds multiplied by 10^6 for legibility. Estimation used semirobust standard errors and was conducted with Stata 11 arima procedure.

Long and short positions for swap dealers and managed money traders, by contrast, are not strongly correlated.

The coefficient ("Coef." in Column 2) shows the marginal (i.e., incremental) effect of independent (control) variables on the dependent variable (here, changes in the log crude oil price). Changes in logs is one way to calculate variables in percentage change terms. For example, the regression results suggest that a 1% change in the Euro/U.S. dollar exchange rate results in a 1.08% increase in the oil price. Results for month variables, included as seasonal controls, and a constant term are not reported. (See text for variable definitions and notes.)

End Notes

[1] James D. Hamilton, "Historical Oil Shocks," University of California, San Diego Department of Economics working paper, February 1, 2011, available at http://dss.ucsd.edu/~jhamilto/oil

[2] Interagency Task Force on Commodity Markets, *Interim Report on Crude Oil*, July 2008, p. 3. Available at http://www.cftc.gov/ucm/groups/public/@newsroom/documents/file/itfinterimreportoncrudeoil0708.pdf.

[3] CFTC Commissioner Bart Chilton, "Speculators and Commodity Prices," Futures Industry Association's Panel Discussion: Financial Investors' Impact on Commodity Prices, Boca Raton, FL, March 16, 2011, available at *http://www.cftc.gov/PressRoom/SpeechesTestimony/opachilton-41.html.*

[4] Kenneth J. Singleton, "Investor Flows and the 2008 Boom/Bust in Oil Prices," March 23, 2011, available at http://www.stanford.edu/~kenneths/OilPub.pdf.

[5] Craig Pirrong, "About Those CFTC Claims That Speculation Has Distorted Oil Prices," Seeking Alpha blog, April 11, 2011, available at http://seekingalpha.com/article/262800-about-those-cftc-claims-that-speculation-

[6] Lutz Kilian and Dan Murphy, "The Role of Inventories and Speculative Trading in the Global Market for Crude Oil," CEPR (Center for Economic and Policy Research) Discussion Papers 7753, May 2010.

[7] U.S. Senate, Committee on Homeland Security and Governmental Affairs, Permanent Subcommittee on Investigations, *Excessive Speculation in the Wheat Market*, Majority and Minority Staff Report, June 24, 2009, available at *http://hsgac.senate.gov/public/_files/*REPORTExcessiveSpecullationintheWheatMarketwoexhibitschartsJune2409.pdf.

[8] U.S. Senate, Committee on Homeland Security and Governmental Affairs, Permanent Subcommittee on Investigations, *Excessive Speculation in the Natural Gas Market*, Staff Report with Additional Minority Staff Views, June 25, 2007, available at http://hsgac.senate.gov/public/index.cfm?FuseAction=Subcommittees.Investigations

[9] *Real Help for American Consumers: Who's Profiting at the Pump?* May 23, 2011, p. 13. Available at http://democrats COOGR%20Democratic%20Oil%20Report%2005-23-1 1 .pdf.

[10] Office of Senator Maria Cantwell, "FTC Launches Investigation into Sky-High Gas Prices After Cantwell's Request," press release, June 20, 2011, available at http://cantwell.senate.gov/news/record.cfm?id=333252. The letter from FTC Chair Jon Leibowitz, quoted in the release, noted that the

Commission has opened an investigation and has authorized the use of compulsory process to determine whether certain oil producers, refiners, transporters, marketers, physical or financial traders, or others (1) have engaged or are engaging in practices that have lessened or may lessen competition – or have engaged or are engaging in manipulation – in the production, refining, transportation, distribution, or wholesale supply of crude oil or petroleum products; or (2) have provided false or misleading information related to the wholesale price of crude oil or petroleum products to a federal department or agency. The Commission seeks to determine through this investigation whether there is a reason to believe that the foregoing practices are in violation of Section 5 of the Federal Trade Commission Act, 15 U.S.C. § 45, as amended; the Commission's Prohibition of Energy Market Manipulation Rule; 16 C.F.R. Part 317; or Section 811 or Section 812 of the Energy Independence and Security Act of 2007, 42 U.S.C. §§ 17301, 17302.

[11] For a recent summary, see Federal Trade Commission, *Report on Activities in the Oil and Natural Gas Industries to the Congressional Appropriations Committees*, July 2010, available at *http://www.ftc.gov/os/2010/07/* P082108oilgasreport.pdf.

[12] Federal Trade Commission, "Prohibitions on Market Manipulation," 74 *Federal Register* 40686-40704, August 12, 2009. For an explanation of those regulations, see Federal Trade Commission, *Guide to Complying with Petroleum Market Manipulation Regulations*, November 2009, available at http://www.ftc.gov/os/2009/11/091113mmrguide.pdf.

[13] For a more detailed exposition of the mechanics of futures, options, and swaps, see CRS Report R40646, *Derivatives Regulation and Recent Legislation*, by Mark Jickling and Rena S. Miller.

[14] An option on a future gives the buyer the right to enter into a long or short futures contract at any time during the life of the option. The option will be exercised only if prices move in favor of the option buyer. The option seller receives a premium in exchange for this right, and profits if the option is not exercised.

[15] Thus, purchase of a futures contract does not mean that 1,000 barrels of physical oil are set aside for eventual delivery. There is no relationship between the volume of futures contracts traded and the amount of oil bought and sold through physical channels. The only significant relationship between the cash and futures market involves price.

[16] In a spread position a trader has a long contract for a given month and a short contract for a different month. In effect, a spread position is a bet that the difference in prices between the futures contracts for the two months will widen or narrow. A spread position, other things equal, carries less risk than an outright long or short position, because whatever happens to the general price level, both legs of the spread will move in tandem to some degree.

[17] Crude oil swaps are not reflected in COT data, because the CFTC receives no information about swap positions. However, once the Dodd-Frank Act's derivatives provisions become effective, swap positions will be disclosed. See CRS Report R41398, *The Dodd-Frank Wall Street Reform and Consumer Protection Act: Title VII, Derivatives*, by Mark Jickling and Kathleen Ann Ruane.

[18] Section 737 of the Dodd-Frank Act (P.L. 111-203) specifies that swap dealers may be treated as bona fide hedgers if they are hedging a trade with a commercial user or producer of a commodity. If they are hedging a trade with a speculator, such as a pension fund, that transaction will be treated as speculation.

[19] CFTC receives information about the identity of traders only when the size of the trader's position exceeds a "reportable" threshold, which for crude oil is 350 contracts.

[20] S. 1200, for example, authorizes the CFTC to impose position limits to reduce the aggregate positions of noncommercial participants and speculators to 35% of the market.

[21] Prior to June 2006, COT data distinguishes only between "commercial" and "non-commercial" traders. The current disaggregation of categories was begun because it was believed that swap dealer positions were mistakenly categorized as "commercial" (i.e., hedging positions).

[22] The oil price figures are Tuesday close-of-market prices, to correspond with COT data, which reflect positions at Tuesday's close.

[23] The same is true of short positions, which are not illustrated here.

[24] Büyükşahin and Robe find that linkages between markets were disrupted in the wake of the Lehman bankruptcy, which among other things, made leveraged finance on which hedge funds rely more costly and difficult. Bahattin Büyükşahin and Michel A. Robe, "Speculators, Commodities and Cross-Market Linkages," American University working paper, February 12, 2011, available at *http://papers.ssrn.com/sol3* /papers.cfm?abstract_id=1707103.

[25] Some caution is warranted when using weekly data to analyze markets that trade in real time. Nevertheless, if longer term trading strategies are most relevant to oil price dynamics, weekly data may capture the most important effects. Büyükşahin and Robe (op. cit.) find no important differences between weekly and daily data in their analysis.

[26] A constant term and indicator variables for months, used for seasonal adjustment, were also included but not reported. Robust standard errors are estimated.

[27] These are not reported here for the sake of brevity, but are available upon request.

[28] In figures presented above this contrast in precision is evident. While the relation between managed-money long positions and price is clear in **Figure 5**, the relation between swap-dealer long positions and price shown in **Figure 7** is less clear.

[29] Holbrook Working, "Futures Trading and Hedging," *American Economic Review*, v. 43, 1953, p. 314.

[30] As **Table 1** shows, they accounted for only 6.6% of total open interest in May 2011.

[31] A similar phenomenon has been observed in the equity market; stocks purchased by investors like Warren Buffett or the Fidelity Magellan Fund under Peter Lynch were almost certain to rise in the short-term because of "copycat" trading.

[32] Firms are said to have market power when they can change the market price of a good or service. In perfectly competitive markets, no firm has market power and all are "price takers" rather than "price makers."

[33] Index investors typically purchase swap contracts that are linked to a commodity price index (reflecting numerous commodities) and gain value when the index rises.

[34] CFTC, *Staff Report on Swap Dealers and Index Traders*, Sept. 11, 2008, p. 39. Available at http://www.cftc.gov/ PressRoom/PressReleases/pr5542-08.html.

[35] Harrison Hong and Jeremy C. Stein, "A Unified Theory of Underreaction, Momentum Trading, and Overreaction in Asset Markets," *Journal of Finance*, vol. 54, no. 6 (Dec. 1999), pp. 2143-2184. Some hedge funds have apparently used more direct means of obtaining information as well. See George Packer, "A Dirty Business: New York City's Top Prosecutor Takes on Wall Street Crime," *New Yorker*, June 27, 2011.

[36] Harrison G. Hong and Motohiro Yogo, "What Does Futures Market Interest Tell Us about the Macroeconomy and Asset Prices?" American Finance Association 2010 Atlanta Meetings Paper, March 13, 2011, available at http://ssrn.com/abstract=1364674.

[37] Utpal Bhattacharya and Xiaoyun Yu, "The Causes and Consequences of Recent Financial Market Bubbles: An Introduction," *Review of Financial Studies*, vol. 21 no. 1(2008), pp. 3-10.

[38] See CRS Report RL33 666, *Asset Bubbles: Economic Effects and Policy Options for the Federal Reserve*, by Marc Labonte.

[39] Susan Shultz, "Commodity of the Quarter: Onions," *Journal of Agricultural and Food Information*, v. 11, 2010, pp. 8-15. The article presents "anecdotal evidence" that some onion farmers would welcome a return of the futures market.

In: U.S. Oil Sources
Editor: Natalie B. Kelleher

ISBN: 978-1-62100-415-8

Chapter 7

THE STRATEGIC PETROLEUM RESERVE AND REFINED PRODUCT RESERVES: AUTHORIZATION AND DRAWDOWN POLICY

Anthony Andrews and Robert Pirog

SUMMARY

Congress authorized the Strategic Petroleum Reserve (SPR) in the Energy Policy and Conservation Act (EPCA, P.L. 94-163) to help prevent a repetition of the economic dislocation caused by the 1973-1974 Arab oil embargo. The Department of Energy (DOE) manages the SPR, which comprises five underground storage facilities, solution-mined from naturally occurring salt domes in Texas and Louisiana. The Energy Policy Act of 2005 (EPAct) authorized SPR expansion to a capacity of 1 billion barrels, but physical expansion of the SPR has not proceeded beyond 727 million barrels — its inventory at the end of 2010. In addition, a Northeast Home Heating Oil Reserve (NHOR) holds 2 million barrels of heating oil in above-ground storage.

EPCA authorized drawdown of the Reserve upon a finding by the President that there is a "severe energy supply interruption." Congress enacted additional authority in 1990 (Energy Policy and Conservation Act Amendments of 1990, P.L. 101-383) to permit use of the SPR for short periods to resolve supply interruptions stemming from situations internal to the United

States. The meaning of a "severe energy supply interruption" has been controversial. EPCA intended use of the SPR only to ameliorate discernible physical shortages of crude oil.

The government had ended the practice of purchasing crude oil to fill the SPR in 1994. In 2000, the Department of Energy began acquiring SPR oil through royalty-in-kind (RIK) in lieu of cash royalties paid on production from federal offshore leases. In May 2008, Congress passed legislation (P.L. 110-232) ordering DOE to suspend RIK fill for the balance of the calendar year unless the price of crude oil dropped below $75/barrel. Crude oil prices spiked to $147/barrel in the summer of 2008 and then sharply declined, allowing a resumption of fill. These activities have brought the SPR essentially to its current 727 million barrel inventory. The current Secretary of the Interior recently announced his intention to terminate the RIK program.

Congress approved $205 million for the SPR in FY2009, including $31.5 million to continue SPR physical expansion activities. DOE has evaluated a site in Richton, MS, as a possible location for an additional 160 million barrels of capacity, but set aside any further expansion plans. The FY2010 Energy and Water Appropriations Act (P.L. 111-85), which provides $243.8 million for the entire SPR program, included $25 million for expansion activities and $43.5 million for purchase of a cavern at Bayou Choctaw to replace a cavern posing environmental risks. The act also prohibits SPR appropriations from being expended to anyone engaged in providing refined product to Iran, or assisting Iran in developing additional internal capacity to refine oil.

Historically, the use of the SPR has been tied to a physical supply shortage, which normally would manifest itself, in part, as a price increase. The original intention of the SPR was to create a reserve of crude oil stocks that could be tapped in the event of an interruption in crude supply. However, price was deliberately kept out of the President's SPR drawdown authority because of concerns about what price level would trigger a drawdown, and that any hint of a price threshold could influence private sector and industry inventory practices. The original intention of the SPR was to create a reserve of crude oil stocks that could be tapped in the event of an interruption in crude supply.

The Government Accountability Office recently observed that the proportion of crude oil grades in the SPR has been growing less compatible with the heavier grades of crude oil that U.S. refineries have been upgrading to handle. This finding has raised questions about the SPR's effectiveness during a long-term oil disruption involving heavy oil.

INTRODUCTION

The recent unrest in the Middle East and North Africa (MENA), disruption in Libyan crude oil production, and rising crude oil prices have led to calls for releasing oil from the Strategic Petroleum Reserve (SPR). In late February, some House members advocated a release as a move to prevent price speculation.[1] More recently, some Senate members formally wrote to the President to urge his exercise of emergency powers to release SPR oil.[2] The Obama Administration is currently considering releasing SPR oil to help ease soaring oil prices and consumer fears over rising gasoline prices. Despite the rising prices, considerable spare capacity and substantial reserves exist worldwide. Up until the last quarter of 2010, the United States had experienced both a drop in crude oil demand and reduced refining output. Auctioning SPR oil may have a limited influence on crude oil prices and find less-than-enthusiastic bidders.

ESTABLISHMENT OF THE SPR

From the mid-1970s through the present day, the United States has had to absorb a number of significant spikes in the price of crude oil and petroleum products.[3] Whether driven by disruptions in the physical supply of crude or refined fuels, or by uncertainties owing to international conflicts and instabilities, these price increases have consequences for the United States. Elevated petroleum prices affect the balance of trade and siphon away disposable income that might be spent to support spending, investment, or savings.

The U.S. Strategic Petroleum Reserve (SPR) originated in the disruption of the 1973 Arab-Israeli War. In response to the United States' support for Israel, the Organization of Arab Exporting Countries (OAPEC) imposed an oil embargo on the United States, the Netherlands, and Canada, and reduced production. While some Arab crude did reach the United States, the price of imported crude oil rose from roughly \$4/barrel (bbl) during the last quarter of 1973 to an average price of \$12.50/bbl in 1974. While no amount of strategic stocks can insulate any oil-consuming nation from paying the market price for oil in a supply emergency, the availability of strategic stocks can help blunt the magnitude of the market's reaction to a crisis. One of the original perceptions of the value of a strategic stockpile was also that its very existence would discourage the use of oil as a political weapon. The embargo imposed by the Arab producers was intended to create a very discernible physical disruption. This explains, in part, why the genesis of the SPR focused especially on deliberate and dramatic physical disruptions of oil flow, and on blunting the significant economic impacts of a shortage stemming from international events.

In response to the experience of the embargo, Congress authorized the Strategic Petroleum Reserve in the Energy Policy and Conservation Act (EPCA, P.L. 94-163) to help prevent a repetition of the economic dislocation caused by the Arab oil embargo. In the event of an interruption, introduction into the market of oil from the Reserve was expected to help calm markets, mitigate sharp price spikes, and reduce the economic dislocation that had accompanied the 1973 disruption. In so doing, the Reserve would also buy time for the crisis to sort itself out or for diplomacy to seek some resolution before a potentially severe oil shortage escalated the crisis beyond diplomacy. The SPR was to contain enough crude oil to replace imports for 90 days, with a goal initially of 500 million barrels in storage. In May 1978, plans for a 750-million-barrel Reserve were implemented. Later authorization expanded it to 1 billion barrels, and the George W. Bush Administration would have expanded it to 1.5 billion barrels had it been successful in persuading Congress of the need.

The Department of Energy (DOE) currently manages the program. Physically, the SPR comprises five underground storage facilities, hollowed out from naturally occurring salt domes, located in Texas and Louisiana. The caverns were finished by injecting water and removing the brine. Similarly, oil is removed by displacing it with water injection. For this reason, crude stored in the SPR remains undisturbed, except in the event of a sale or exchange. Multiple injections of water, over time, will compromise the structural integrity of the caverns.[4] By 2005, the storage capacity of the SPR expanded to 727 million barrels, and its inventory had reached nearly 700 million barrels before Hurricanes Katrina and Rita in 2005. Following the storms, some crude was loaned to refiners and some was sold. Loans of SPR oil are "paid" by the return of larger amounts of oil than were borrowed.[5] The SPR has since been filled to its 727 million barrel capacity through royalty-in-kind acquisition, which this report discusses further below.

SPR oil is sold competitively. A "notice of sale" is issued, including the volume, characteristics, and location of the petroleum for sale; delivery dates and procedures for submitting offers; as well as measures for assuring performance and financial responsibility. Bids are reviewed by DOE and awards offered. The Department of Energy estimates that oil could enter the market roughly two weeks after the appearance of a notice of sale.[6]

The SPR could be drawn down initially at a rate of roughly 4.4 million barrels per day for up to 90 days; thereafter, the rate would begin to decline. Although fears were expressed periodically during the 1980s about whether the facilities for withdrawing oil from the Reserve were in proper readiness, the absence of problems during the first real drawdown in early 1991 (the Persian Gulf War) appeared to allay much of that concern. However, some SPR facilities and infrastructure were beginning to reach the end of their operational life. A life extension program, initiated in 1993, upgraded or replaced all major systems to ensure the SPR's readiness to 2025.

The Arab oil embargo also fostered the establishment of the International Energy Agency (IEA) to develop plans and measures for emergency responses to energy crises. Strategic stocks are one of the policies included in the agency's International Energy Program (IEP). Signatories to the IEA[7] are committed to maintaining emergency reserves representing 90 days of net imports, developing programs for demand restraint in the event of emergencies, and agreeing to participate in allocation of oil deliveries among the signatory nations to balance a shortage among IEA members. The calculation of net imports for measuring compliance with the IEA requirement includes private stocks. By that measure, the United States has more than 100 days' cushion. However, it is likely that less than 20% of the privately held stocks would technically be available in an emergency, because most of that inventory supports movement of product through the delivery infrastructure. At full capacity, the SPR might afford the United States roughly 70 days or more of net import protection, depending upon the pace of recovery of the domestic economy. These measures of days of protection assume a total cessation of oil supply to importing nations, a scenario that is highly unlikely. This would be especially true for the United States, given that Canada is currently the nation's principal source for crude oil.

Some IEA member nations require a level of stocks to be held by the private sector or by both the public and private sectors. Including the U.S. SPR, roughly two-thirds of IEA stocks are held by the oil industry, whereas one-third is held by governments and supervisory agencies.[8]

SPR DRAWDOWN AUTHORITIES

The Energy Policy and Conservation Act (EPCA, P.L. 94-163) authorized drawdown of the Reserve upon a finding by the President that there is a "severe energy supply interruption." This was deemed by the statute to exist if three conditions were joined: If "(a) an emergency situation exists and there is a significant reduction in supply which is of significant scope and duration; (b) a severe increase in the price of petroleum products has resulted from such emergency situation; and (c) such price increase is likely to cause a major adverse impact on the national economy."

Congress enacted additional drawdown authority in 1990 (Energy Policy and Conservation Act Amendments of 1990, P.L. 101-383) after the Exxon Valdez oil spill, which interrupted the shipment of Alaskan oil, triggering spot shortages and price increases. The intention was to provide for an SPR drawdown under a less rigorous finding than that mandated by EPCA. This section, 42 U.S.C. § 6241(h), has allowed the President to use the SPR for a short period without having to declare the existence of a "severe energy supply interruption" or the need to meet obligations of the United States under the international energy program. The Energy Policy Act of 2005 made the SPR authorities permanent. These authorities also provided for U.S. participation in emergency-sharing activities of the International Energy Agency without risking violation of antitrust law and regulation.

Under the additional authorities authorized in P.L. 101-383, a drawdown may be initiated in the event of a circumstance that "constitutes, or is likely to become, a domestic or international energy supply shortage of significant scope or duration" and where "action taken ... would assist directly and significantly in preventing or reducing the adverse impact of such shortage." This authority allows for a limited use of the SPR. No more than 30 million barrels may be sold over a maximum period of 60 days, and this limited authority may not be exercised at all if the level of the SPR is below 500 million barrels. This was the authority behind the Bush Administration's offer of 30 million barrels of SPR oil on September 2, 2005, which was part of the coordinated drawdown called for by the International Energy Agency. The same authority may have been the model for a swap ordered by President Clinton on September 22, 2000.

Acquisition of Crude Oil for the SPR

From 1995 until the latter part of 1998, sales of SPR oil, not acquisition, were at the center of debate. However, the subsequent reduction and brief elimination of the annual federal budget deficit—as well as a precipitous drop in crude oil prices into early 1999—generated new interest in replenishing the SPR, either to further energy security objectives or as a means of providing price support to domestic producers who were struggling to keep higher-cost, marginal production in service.

Royalty-in-Kind Acquisition

As an alternative to appropriations for the purchase of SPR oil, DOE proposed that a portion of the royalties paid to the government from oil leases in the Gulf of Mexico be accepted "in kind" (in the form of oil) rather than as revenues. The Department of the Interior (DOI) was reported to be unfavorably disposed to the royalty-in-kind (RIK) proposal, but a plan to proceed with such an arrangement was announced on February 11, 1999. (Legislation had also been introduced [H.R. 498] in the 106th Congress to direct the Minerals Management Service to accept royalty-in-kind oil.) Producers were supportive, maintaining that the system for valuation of oil at the wellhead is complex and flawed. While acquiring oil for the SPR by RIK avoids the necessity for Congress to make outlays to finance direct purchase of oil, it also means a loss of revenues in so far as the royalties are settled in wet barrels rather than paid to the U.S. Treasury in cash. Final details were worked out during the late winter of 1999.

In mid-November of 2001, President Bush ordered fill of the SPR to 700 million barrels, principally through oil acquired as royalty-in-kind (RIK). At its inception, the RIK plan was generally greeted as a well-intended first step toward filling the SPR to its capacity of 727 million barrels.[9] However, it became controversial when crude prices began to rise sharply in 2002. Some policymakers and studies asserted that diverting RIK oil to the SPR instead of selling it in the open market was putting additional pressure on crude prices. A number of industry analysts argued that the quantity of SPR fill was not enough to have driven the market. The Administration strongly disagreed with claims that RIK fill bore responsibility for the continuing spike in prices.

Legislative attempts to suspend RIK fill began in 2004, during the 108th Congress. The Energy Policy Act of 2005 (P.L. 109-58), enacted in the summer of 2005, required the Secretary of Energy to develop and publish for comment procedures for filling the SPR that take into consideration a number of factors. Among these are the loss of revenue to the Treasury from accepting royalties in the form of crude oil, how the resumed fill might affect prices of both crude and products, and whether additional fill would be justified by national security. On November 8, 2006, DOE issued its final rule, "Procedures for the Acquisition of Petroleum for the Strategic Petroleum Reserve." The rule essentially indicated that DOE would take into account all the parameters required by P.L. 109-58 to be taken into consideration before moving ahead with any acquisition strategy. DOE rejected tying decisions to acquire oil to any specific, measurable differentials in current and historic oil prices.

In the summer of 2007, DOE resumed RIK fill of the SPR. On May 19, 2008, with gasoline prices exceeding, on average, $3.60 gallon, and approaching $4.00/gallon in some regions, Congress passed P.L. 110-232. However, a few days earlier, on May 16, DOE announced it would not accept bids for an additional 13 million barrels of RIK oil that had been intended for delivery during the second half of 2008.

Through FY2007, royalty-in-kind deliveries to the SPR totaled roughly 140 million barrels and forgone receipts to the Department of the Interior an estimated $4.6 billion. DOE had estimated deliveries of 19.1 million barrels of RIK oil during FY2008 and $1.170 billion in forgone revenues.[10]

Opponents of RIK fill in the 110th Congress were not necessarily opposed to the concept of an SPR. When the price of crude was much less of an issue, objections to RIK fill were also ideological. Opponents of RIK fill in principle contended that a government-owned strategic stock of petroleum was inappropriate under any circumstance—that it essentially saddled the public sector with the expense of acquiring and holding stocks, the cost for which might have otherwise been borne by the private sector. The existence of the SPR, this argument goes, has blunted the level of stocks held in the private sector.[11] As already noted, RIK fill resumed in 2009 and ended in early 2010, pending establishment of additional storage. A site in Richton, MS, has been evaluated as a possible site for expansion of the SPR. However, while $25 million for expansion activities was included in the FY2010 budget, it is not apparent that expansion is a high priority.

Royalty-in-Kind Termination

As already noted, legislation (P.L. 110-232) enacted in May 2008 forbade DOE from initiating any new activities to acquire royalty-in-kind (RIK) oil for the SPR during the balance of 2008. The sharp decline in crude oil prices since spiking to $147/barrel in the summer of 2008 had spurred interest in resuming fill of the SPR. On January 2, 2009, the Bush Administration announced plans to purchase oil for the SPR, and to reschedule deferred deliveries. There were four components in the resumption of fill: (1) a purchase announced on January 16, 2009, of nearly 10.7 million barrels to replace oil that was sold after Hurricanes Katrina and Rita in 2005; (2) the return of roughly 5.4 million barrels of oil borrowed by refiners after Hurricane Gustav in 2008; (3) delivery of roughly 2.2 million barrels of RIK oil that had been deferred; and (4) resumption of RIK fill in May 2009 at a volume of 26,000 barrels per day, totaling over 6.1 million barrels to be delivered over a period from May 2009 to January 2010. These activities were intended to fill the SPR to its current capacity of 727 million barrels by early 2010. The government has not acquired oil for the SPR by outright purchase since 1994, when oil purchases ended. The SPR then held 592 million barrels.

On September 16, 2009, Secretary of the Interior Ken Salazar announced a transitional phasing out of the RIK Program.[12] As RIK oil and natural gas sales contracts expire, the oil and natural gas properties will revert to in-value status. As a result of the decision, an expansion of the SPR would require purchasing oil through the market at prevailing prices.

The Northeast Home Heating Oil Reserve

The Northeast Home Heating Oil Reserve (NHOR) was established after a number of factors contributed to the virtual doubling in some Northeastern locales of home heating oil prices during the winter of 1999-2000. Drawing particular attention of lawmakers was the sharply lower level of middle distillate stocks—from which both home heating oil and diesel fuels are produced—immediately beforehand. EPCA includes authority for the Secretary of Energy to establish regional reserves as part of the broader Strategic Petroleum Reserve. With support from the Clinton Administration, Congress moved to specifically authorize and fund a regional heating oil reserve in the Northeast. The FY2001 Interior Appropriations Act (P.L. 106-291) provided $8 million for the Northeast Home Heating Oil Reserve

(NHOR). The regional reserve was filled by the middle of October 2000 at two sites in New Haven, CT, and terminals in Woodbridge, NJ, and Providence, RI. The NHOR is intended to provide roughly 10 days of Northeast home heating oil demand.

There was controversy over the language that would govern its use. Opponents of establishing a regional reserve suspected that it might be tapped at times that some consider inappropriate, and that the potential availability of the reserve could be a disincentive for the private sector to maintain inventories as aggressively as it would if there were no reserve. The approach enacted predicated drawdown on a regional supply shortage of "significant scope and duration," or if—for seven consecutive days—the price differential between crude oil and home heating oil increased by more than 60% over its five-year rolling average. The intention was to make the threshold for use of the regional reserve high enough so that it would not discourage oil marketers and distributors from stockbuilding. The President could also authorize a release of the NHOR in the event that a "circumstance exists (other than the defined dislocation) that is a regional supply shortage of significant scope and duration," the adverse impacts of which would be "significantly" reduced by use of the NHOR.

During mid- and late December 2000, the 60% differential was breached. However, this was due to a sharp decline in crude prices rather than to a rise in home heating oil prices. In fact, home heating oil prices were drifting slightly lower during the same reporting period. As a consequence, while the 60% differential was satisfied, other conditions prerequisite to authorizing a drawdown of the NHOR were not.

A general strike in Venezuela that began in late 2002 resulted, for a time, in a loss of as much as 1.5 million barrels of daily crude supply to the United States. With refinery use lower than usual owing to less crude reaching the United States, domestic markets for home heating oil had to rely on refined product inventories to meet demand during a particularly cold winter. Prices rose, and there were calls for use of the NHOR; still, the price of heating oil fell significantly short of meeting the guidelines for a drawdown.[13]

In the 111th Congress, S. 283, introduced on January 21, 2009, would have permitted drawdown on the basis of price as well as supply. The bill would have mandated a release of 20% of the heating oil held in the Reserve if the average retail price for home heating oil in the Northeast exceeded $4.00 per gallon on November 1 of the fiscal year. An additional 20% would be released in four additional installments if the average retail price exceeded $4.00/gallon on the first of each month, December through March. S. 283 was the subject of a May 12, 2009 hearing before the Senate Committee on Energy and Natural Resources (S. Hrg. 111-67).

SPR Expansion

The Energy Policy Act of 2005 (EPAct) requires, "as expeditiously as practicable," expansion of the SPR to its authorized maximum physical capacity of 1 billion barrels. Advocates for expansion argued that the SPR would need to be larger for the United States to be able to maintain stocks equivalent to 90 days of net imports. The FY2010 SPR budget, at $229 million, included $43.5 million for purchase of a cavern at Bayou Choctaw to replace a cavern posing environmental risks, as well as $25 million for expansion activities. However, it is not apparent that expansion remains a high priority.

Bayou Choctaw

Bayou Choctaw has six storage caverns with a total capacity of 76 million barrels. One cavern, Bayou Choctaw Cavern 20, is within 60 ft of the edge of the salt dome and must be replaced for fear of breaching the salt dome with further use.[14] The SPR oil stored at the Bayou Choctaw site is not in any immediate danger of leaking, however. Cavern 20 has passed all integrity tests. To limit any risk, DOE has reduced the oil stored in Cavern 20 from 7.5 million barrels to 3.2 million barrels by using only the upper portion of the cavern.

DOE has temporarily stored the remaining oil in Big Hill and West Hackberry caverns in the brine cushion at the bottom of the cavern, which will eventually be needed to accommodate cavern creep (geological shrinkage). DOE is currently acquiring a 10-million-barrel replacement cavern in the Bayou Choctaw salt dome. The replacement cavern would be available for oil storage in December 2012 and would provide capacity for the 3.2 million barrels in Cavern 20 and the 4.3 million barrels temporarily stored at Big Hill and West Hackberry, plus provide an additional 2.5 million barrels of spare capacity. Cavern 20 is to be emptied and abandoned.[15]

Richton Dome

The Energy Policy Act of 2005 (EPAct) required SPR expansion to its maximum authorized capacity of 1 billion barrels. DOE has evaluated a site in Richton, MS, as a possible location for an additional 160 million barrels of capacity. However, in its FY2011 request, the Obama Administration proposed to suspend spending in support of SPR expansion. The budget request proposed to redirect $71 million previously appropriated for expansion to "partially fund SPR non-expansion operations and maintenance activities." The Administration cited an Energy Information Administration projection that "U.S. petroleum consumption and dependence on imports will decline in the future and the current Reserve's projection will gradually increase to 90 days by 2025." The Administration reduced the FY2011 request for the SPR to $138.9 million, a sharp reduction from the $243.8 million appropriated for FY2010. The FY2012 SPR funding request is reduced further, to $121.7 million.

THE DEBATE OVER WHEN THE SPR SHOULD BE USED

Historically, the use of the SPR has been tied to a physical shortage of supply—which normally will manifest itself, in part, in an increase in price. However, price was deliberately kept out of the President's SPR drawdown authority because of concerns about what price level would trigger a drawdown, and that any hint of a price threshold could influence private sector and industry inventory practices.

As has been noted, the original intention of the SPR was to create a reserve of crude oil stocks that could be tapped in the event of an interruption in crude supply. Debate over releasing SPR oil traces differences in opinion over just what constitutes a "severe energy supply interruption." The debate during the 1980s over when, and for what purpose, to initiate an SPR drawdown reflected the significant shifts that were taking place in the operation of oil markets after the experiences of the 1970s, and deregulation of oil price and supply. Sales of SPR oil authorized by the 104th Congress—and in committee in the 105th—renewed the debate for a time.[16]

The SPR Drawdown Plan, submitted by the Reagan Administration in 1982, provided for price-competitive sale of SPR oil. The plan rejected the idea of conditioning a decision to distribute SPR oil on any "trigger" or formula. To do so, the Administration argued, would discourage private sector initiatives for preparedness or investment in contingency inventories. Many analysts, in and out of Congress, agreed with the Administration that reliance upon the marketplace during the shortages of 1973 and 1979 would probably have been less disruptive than the price and allocation regulations that were imposed. But many argued that the SPR should be used to moderate the price effects that can be triggered by shortages like those of the 1970s or the tight inventories experienced during the spring of 1996, and lack of confidence in supply availability. Early drawdown of the SPR, some argued, was essential to achieve these objectives.

The Reagan Administration revised its position in January 1984, announcing that the SPR would be drawn upon early in a disruption. This new policy was hailed as a significant departure, considerably easing congressional discontent over the Administration's preparedness policy, but it also had international implications. Some analysts began to stress the importance of coordinating stock drawdowns worldwide during an emergency lest stocks drawn down by one nation merely transfer into the stocks of another and defeat the price-stabilizing objectives of a stock drawdown. In July 1984, responding to pressure from the United States, the International Energy Agency agreed "in principle" to an early drawdown, reserving decisions on "timing, magnitude, rate and duration of an appropriate stockdraw" until a specific situation needed to be addressed.

Use of the SPR in the Persian Gulf War (1990)

This debate was revisited in the aftermath of the Iraqi invasion of Kuwait on August 2, 1990. The escalation of gasoline prices and the prospect that there might be a worldwide crude shortfall approaching 4.5 million-5.0 million barrels daily prompted some to call for drawdown of the SPR. The debate focused on whether SPR oil should be used to moderate anticipated price increases, before oil supply problems had become physically evident.

In the days immediately following the Iraqi invasion of Kuwait, the George H. W. Bush Administration indicated that it would not draw down the SPR in the absence of a physical shortage simply to lower prices. On the other hand, some argued that a perceived shortage does as much and more immediate damage than a real one, and that flooding the market with stockpiled oil to calm markets is a desirable end in itself. From this perspective, the best opportunity to use the SPR during the first months of the crisis was squandered. It became clear during the fall of 1990 that in a decontrolled market, physical shortages are less likely to occur. Instead, shortages are likely to be expressed in the form of higher prices, as purchasers are free to bid as high as they wish to secure scarce supply.

Within hours of the first air strike against Iraq in January 1991, the White House announced that President Bush was authorizing a drawdown of the SPR, and the IEA activated the plan on January 17. Crude prices plummeted by nearly $10/barrel in the next day's trading, falling below $20/bbl for the first time since the original invasion. The price drop was attributed to optimistic reports about the allied forces' crippling of Iraqi air power and the diminished likelihood, despite the outbreak of war, of further jeopardy to world oil supply. The IEA plan and the SPR drawdown did not appear to be needed to help settle markets, and there was some criticism of it. Nonetheless, more than 30 million barrels of SPR oil was put out to bid, but DOE accepted bids deemed reasonable for 17.3 million barrels. The oil was sold and delivered in early 1991.

The Persian Gulf War was an important learning experience about ways in which the SPR might be deployed to maximize its usefulness in decontrolled markets. As previously noted, legislation enacted by the 101st Congress, P.L. 101-383, liberalized drawdown authority for the SPR to allow for its use to prevent minor or regional shortages from escalating into larger ones; an example was the shortages on the West Coast and price jump that followed the Alaskan oil spill of March 1989. In the 102nd Congress, omnibus energy legislation (H.R. 776, P.L. 102-486) broadened the drawdown authority further to include instances where a reduction in supply appeared sufficiently severe to bring about an increase in the price of petroleum likely to "cause a major adverse impact on the national economy." The original EPCA authorities permit "exchanges" of oil for the purpose of acquiring additional oil for the SPR. Under an exchange, a company borrows SPR crude and later replaces it, including an additional quantity of oil as a premium for the loan. There were seven exchanges between 1996 and 2005. The most recent one (with the exception of a test exchange in the spring of 2008) was in June 2006. ConocoPhillips and Citgo borrowed 750,000 barrels of sour crude for two refineries affected by temporary closure of a ship channel.

A new dimension of SPR drawdown and sale was introduced by the Clinton Administration's proposal in its FY1996 budget to sell 7 million barrels to help finance the SPR program. While agreeing that a sale of slightly more than 1% of SPR oil was not about to cripple U.S. emergency preparedness, some in Congress vigorously opposed the idea, in part because it might establish a precedent that would bring about additional sales of SPR oil for purely budgetary reasons, as did indeed occur. There were three sales of SPR oil during FY1996. The first was to pay for the decommissioning of the Weeks Island site. The second was for the purpose of reducing the federal budget deficit, and the third was to offset FY1997 appropriations. The total quantity of SPR sold was 28.1 million barrels, and the revenues raised were $544.7 million. Fill of the SPR with RIK oil was initiated in some measure to replace the volume of oil that had been sold during this period.

Hurricanes and Changes in the Market Dynamics (2005-2008)

Prior to Hurricanes Ivan, Katrina, and Rita in 2005, growth in oil demand had begun to strap U.S. refinery capacity. A result has been an altering in a once-observed historic correlation between crude oil and refined oil product prices. In the past, changes in the price of crude had driven changes in the cost of refined products. The assumption that product prices are driven by, and follow the path of, crude prices, was at the center of debates from the 1980s until early in the decade of 2000 whether an SPR drawdown was warranted when prices spiked.

However, beginning in the middle of the first decade of the new century, pressure on product supplies and the accompanying anxiety stoked by international tensions caused a divorce in that traditional correlation between crude and product prices. The increases in prices of gasoline and other petroleum products following Hurricanes Katrina and Rita, for example, were not a response to any shortage of crude oil, but to shortages of refined products owing to the shutdown of major refining capacity in the United States, and to an interruption of product transportation systems.

The rise in crude prices to over $140/barrel by the summer of 2008 was attributable to many contributing factors, including increasing international demand, and concern that demand for crude might outstrip world production. Markets were described as "tight," meaning that there might be little cushion in terms of spare production capacity to replace any crude lost to the market, or to provide adequate supply of petroleum products. In such a market, where demand seems to be brushing against the limits to meet that demand, refinery outages, whether routine or unexpected, can spur a spike in crude and product prices, as can weekly reports of U.S. crude and petroleum stocks, if the numbers reported are not consistent with expectations. As prices continued to increase during 2007-2008, some argued that market conditions did not support the high prices. One market analyst remarked at the end of October

2007, "The market at this stage totally ignores any bearish news [that would soften the price of oil], but it tends to exaggerate bullish news."[17] Significant and sustained increases in oil prices were observed in the absence of the sort of "severe energy supply interruption" that remains the basis for use of the SPR. Legislation in the 111th Congress (S. 1462) would have established a price basis for authorizing a drawdown of the SPR. However, a release from the SPR might not lower prices under every scenario.

Summer 2008 Call for an SPR Drawdown

Some policymakers had urged the George W. Bush Administration to release oil from the SPR during the crude oil price run-up of spring and summer 2008. A review of the dynamics in the oil market during this period provides a demonstration of why an SPR release in the face of high prices will not necessarily foster a decline in petroleum prices.

By mid-July 2008, U.S. gasoline prices were exceeding $4.00/gallon and diesel fuel was averaging $4.75/gallon. Crude oil prices had briefly exceeded more than $145/barrel, but declined late in the month to less than $128/barrel. Oil prices had risen in recent years in the absence of the normal association with the concept of "disruption" or "shortage." The escalation in prices to their observed peak in July 2008 was driven by several factors that are difficult to weigh. Chief among them was the existence of little or no spare oil production capacity worldwide, and a general inelasticity in demand for oil products despite high prices. Prices also generally prove sensitive to the ebb and flow of international tensions, the value of the U.S. dollar, and even the appearance of storms that could develop into hurricanes that might make landfall in the Gulf of Mexico.

Political Unrest in the Middle East and North Africa (MENA) 2011

The political unrest that began in Tunisia and spread to Egypt and Libya in early 2011 has been tied to the surge in oil prices observed during the 1st quarter of the year. To calm markets and to moderate prices, some have called for releasing oil from the SPR.

By early March 2011, the price of West Texas Intermediate (WTI), trading on the New York Mercantile Exchange (NYMEX) for April delivery, was in excess of $100 per barrel. In Europe, the price of Brent crude oil exceeded $115 per barrel. These prices, approximately 20% higher than before the outbreak of political unrest, reflect at least two important factors: first, expectations that the unrest could spread to other countries, some of which could be major oil producers, and second, that actual Libyan exports have been curtailed, to an uncertain extent, but likely by at least one half. As an offset to the lost Libyan crude exports, Saudi Arabia has indicated that it will expand its exports to keep the world market supplied.

If oil were released from the SPR, the key questions would be whether such a release could dampen expectations that are driving the market, and whether it would reduce prices in the short run, and where.

The world is not currently experiencing a shortage of crude oil. Demand remains moderated by the recession, although it is recovering. On the supply side, Saudi Arabia, it is believed, has expanded production to replace the Libyan shortfall. However, a shortage does exist in the sense that Libyan exports of light, sweet crude oil, favored by European refiners to produce clean diesel fuel, have fallen. It is not likely that heavier, sour crude oil produced by Saudi Arabia is a perfect substitute for lost Libyan supplies. As a result, European prices of crude oil have risen more than U.S. prices. If the United States drew light, sweet crude oil from the SPR in sufficient volume, it is possible that competition between U.S. and European refiners for available oil supplies could moderate competing demands and have some effect on price. Since prices in Europe rose the most due to unrest in the MENA, it is likely that any moderating in price due to a release of light, sweet crude from the SPR would occur in the European market with only a smaller effect in the United States. It is not known what volume of SPR draw would be required for any specified price reduction, and it is worth noting that of the 727 million barrels in the SPR, less than half, about 293 million barrels, is light, sweet crude.

If political unrest does spread to additional major oil exporting countries, for example Kuwait or Saudi Arabia, it is likely that a major increase in the world price of oil, based on deteriorating expectations, would occur. In that case, additional available oil on the market from reserve stocks would likely have little effect on futures markets, as price movements in those markets reflect availability of supply into the future, not current market conditions. Rather than more currently available oil, the futures markets would require a belief that more future oil production capacity was likely to become available.

OTHER POLICY CONSIDERATIONS

Over the last 25 years, the °API gravity of imported crude oils has been decreasing, while average sulfur content has been increasing. °API gravity, a measure developed by the American Petroleum Institute, expresses the "lightness" or "heaviness" of crude oils on an inverted scale.[18] With a diminishing supply of light, sweet (low sulfur) crude oil, U.S. refineries have had to invest in multi-billion dollar processing-upgrades to convert lower-priced heavier, sour crude oils to high-value products such as gasoline, diesel, and jet fuel.

Currently, 124 U.S. refineries process crude oil into fuel (this includes three refinery complexes that are each made up of two formerly independent refineries).[19] The number is down from the 158 reportedly operating a decade ago. Although the number of refineries has decreased, operable refining capacity has increased over the past decade from 16.5 million barrels/day (bpd) to over 18 million barrels per day. Most of the country's gasoline is refined in the Gulf Coast region (Petroleum Administration for Defense District 3), which makes up nearly 45% of the U.S. refining capacity through 45 refineries processing more than 8 million barrels per day. These refineries also represent some of the largest and most complex refineries in the United States, if not the world.

In the months prior to Hurricanes Gustav and Ike, there were some calls for an SPR drawdown despite the absence of any discernible shortage. On July 24, 2008, legislation (H.R. 6578; 110th Congress) to require a 10% drawdown of SPR oil failed to achieve a two-thirds majority in the House under suspension of the rules (226-190). The language was included in H.R. 6899 (110th Congress), the Comprehensive American Energy Security and Consumer Protection Act, which passed the House on September 16th (236-189). The bill would have required a sale of 70 million barrels of light grade petroleum from the SPR within six months following enactment. The bill stipulated that 20 million barrels must be offered for sale during the first 60 days. All oil from the sale would be replaced with "sour" crude to be acquired after the six-month sale period, with the replacement acquisition completed not later than five years after enactment.

The genesis of the H.R. 6899 proposal lay partly in an analysis by the Government Accountability Office (GAO), which observed that the proportion of grades of oil in the SPR was not as compatible as it could be with the trend of refineries toward being able to handle heavier grades of crudes.[20] GAO observed that 40% of the crude oil refined by U.S. refineries was heavier than that stored in the SPR. Refiners reported to GAO that running lighter crude in units designed to handle heavy crudes could impose as much as an 11% penalty in gasoline production and 35% in diesel production. The agency reported that other refiners indicated that they might have to shut down some of their units.[21] Refineries that process heavy oil cannot operate at normal capacity if they run lighter oils. The types of oil currently stored in the SPR would not be fully compatible with 36 of the 74 refineries considered vulnerable to supply disruptions. (A majority of the refineries that have pipeline access to the SPR are located in the Gulf Coast region and the Midwest region.) GAO cited a DOE estimate that U.S. refining throughput would decrease by 735,000 barrels per day (or 5%) if the 36 refineries had to use SPR oil—a substantial reduction in the SPR's effectiveness during an oil disruption, especially if the disruption involved heavy oil.

It was unclear what sort of effect a roughly 70 million barrel draw on the SPR would have on prices. In a market where there is no physical shortage, oil companies may have limited interest in SPR oil unless they have spare refining capacity to turn the crude into useful products, or want to build crude oil stocks.[22] SPR oil is not sold at below-market prices. Bids on SPR oil are accepted only if the bids are deemed fair to the U.S. government. If the announcement itself that the SPR is going to be tapped does not prompt or contribute to a softening of prices, there may be limited interest on the part of the oil industry in bidding on SPR supply. Although the possibility exists that prices might decline if additional refined product is released into the market, it was impossible to predict what effect an SPR drawdown would have had on oil prices at any time in 2008, given the many other factors that bear on daily oil prices.

There are additional considerations. A unilateral draw on U.S. stocks would probably have less impact on the world oil market than a coordinated international drawdown of the sort that occurred after Hurricanes Katrina and Rita in 2005. Some might argue that it would be unwise under any scenario for the United States to draw down its strategic stocks while other nations continue to hold theirs at current levels. Additionally, it is always possible that producing nations might reduce production to offset any SPR oil delivered into the market. In the setting of 2008, producing, exporting nations could have argued that the market was already well-supplied and that short-term supply concerns were not what was keeping prices relatively high.

The SPR has been perceived as a defensive policy tool against high oil prices, but if it is used without a discernible impact on oil prices, it is possible that the SPR will lose some of whatever psychological leverage it exercises on prices when left as an untapped option.

LEGISLATION IN THE 111TH CONGRESS

The Omnibus Appropriations Acts of 2009 (P.L. 111-8) provided $205 million for necessary expenses for Strategic Petroleum Reserve facility development and operations and program management activities, of which $31.5 million was provided to initiate new site expansion activities, beyond land acquisition. The Northeast Home Heating Oil Reserve received $9.8 million for necessary operation and management expenses.

The Supplemental Appropriations Act for 2009 (P.L. 111-32) authorized a transfer of $21.6 million from the SPR petroleum account for site maintenance activities.

The Energy and Water Appropriations Act for 2010 (P.L. 111-85) provided $243.8 million for development and operations and program management activities at the Strategic Petroleum Reserve facility, and $11.3 million for the Northeast Home Heating Oil Reserve. Section 313 of the Appropriations Act prohibited expending SPR appropriations to anyone engaged in selling refined products valued at more than $1 million to Iran or contributing in any way to expansion of refining capacity in Iran. Firms providing, or insuring tankers carrying, refined product to Iran were also included in the prohibition.

PROPOSALS IN THE 112TH CONGRESS

A bill to establish a National Strategic Gasoline Reserve (H.R. 142) would authorize the Secretary of Energy to set aside 10 million barrels of refined gasoline products similar to the Northeast Home Heating Oil Reserve.

Arguments in favor of establishing a refined product reserve are that U.S. oil imports include refined products and that it could be more efficient and calming to markets if it were not necessary to first draw down SPR crude and then refine it into needed products. The effect that SPR crude might have on moderating price increases could also be offset if refineries themselves or oil pipelines carrying crude to refineries were compromised. The availability of refined product reserves would address that scenario. Having a regional product reserve would also lessen the likelihood that delivery of crude or product from the stocks of IEA signatories might overwhelm U.S. port facilities; this happened in the wake of the European response that followed Hurricanes Rita and Katrina.

Arguments against a product reserve include the prospect that the availability of supplemental supplies of gasoline from abroad may increase as European demand for diesel vehicles displaces gasoline consumption there. Additionally, storage of refined product is more expensive than for crude. Storage of crude in salt caverns is estimated to cost roughly $3.50/barrel while above-ground storage of product in tanks might cost $15-$18/barrel. Refined product will also deteriorate and would need to be periodically sold and replaced to assure the quality of the product held in the product reserve. Many states also use different gasoline blends, adding to the complexity of identifying which blends should be stored where, and in what volume. It would be simpler to hold conventional gasoline in a product reserve with the expectation that the Environmental Protection Agency (EPA) would waive Clean Air Act (CAA) requirements during an emergency.[23]

H.R. 1017, introduced March 10, 2011, would provide for the sale of light grade petroleum from the Strategic Petroleum Reserve and replace the crude with refined petroleum product.

End Notes

[1] George Lobsenz, "House Democrats to Obama: Tap SPR to Stop Speculators," *The Energy Daily*, February 25, 2011, p. 1, Vol . 39, No. 38.

[2] "Senators Ask Obama for Swift Action to Lower Gas Prices," March 4, 2011, Senator Jeff Merkley, http://merkley.senate.gov/newsroom/press/release/?id=F3401222-9994-455F-9BA9-BFD604ADCAF0.

[3] These have included the Arab oil embargo (1973-1974), the deposing of the Shah of Iran, followed by the Iranian revolution (1979-1980), the first Gulf War (1990), and production cuts by the Organization of the Petroleum Exporting Countries (OPEC) and a resurgence in world oil demand (early 1999 into the fall of 2000). Starting in 2003 with the U.S. invasion of Iraq, crude oil and product prices began rising to a new nominal high reached in the summer of 2008—owing to a blend of speculation in the futures market and adjustment in the real buying power of the U.S. dollar. Some of the dynamics behind more recent prices are linked to natural events (such as Hurricanes Rita, Katrina, and Gustav) that disrupted Gulf Outer Continental Shelf (OCS) oil production and the 2010 BP Macondo well blowout that resulted in a moratorium on OCS drilling permits. The latest rise, at the time of this paper, may depend on how events in Libya play out.

[4] Oil stored at one SPR site, Weeks Island, was transferred after problems with the structural integrity of the cavern—unrelated to drawdown activity—were discovered in the mid-1990s.

[5] Details and current levels of SPR inventory are updated regularly at http://www2.spr.doe.gov/DIR/SilverStream/ Pages/pgDailyInventoryReportViewDOE_new.html.

[6] See http://www.fe.doe.gov/programs/reserves/spr/spr-facts.html. For more detail on the sales procedure, see U.S. *Federal Register*, Department of Energy, *Price Competitive Sale of Strategic Petroleum Reserve Petroleum; Standard Sales Provisions: Final Rule*, July 27, 2005, pp. 39363-39382; available at http://www.fe.doe.gov/programs/reserves/ spr/spr_rule_070705.pdf. The Department of Energy has a history of SPR drawdowns, sales, and exchanges on the web at http://www.fe.doe.gov/programs/reserves/spr/spr-drawdown.html.

[7] IEA member countries are Australia, Austria, Belgium, Canada, Czech Republic, Denmark, Finland, France, Germany, Greece, Hungary, Ireland, Italy, Japan, Republic of Korea, Luxembourg, the Netherlands, New Zealand, Norway, Portugal, Spain, Sweden, Switzerland, Turkey, United Kingdom, and the United States. See http://www.iea.org/ Textbase/about/membercountries.asp.

[8] See http://www.iea.org/Textbase/subjectqueries/keyresult.asp?KEYWORD_ID=4103.

[9] The SPR estimated capacity of 727 million barrels followed a reevaluation of the cavern formations and other work. Water injections into caverns when oil has been moved have added capacity, as did completion of a project to remove excess gas from stored petroleum.

[10] Owing to suspension of RIK fill after the passage of legislation in May 2008, these figures will be significantly lower. Annual figures for RIK deliveries through FY2006 may be found in the Strategic Petroleum Reserve Annual Report for FY2006, p. 39: http://www.fossil.energy.gov/programs/reserves/publications/Pubs-SPR/ spr_annual_rpt_06.pdf. Estimates for FY2008 furnished in a communication from DOE.

[11] See, for example, Taylor, Jerry and Van Doren, Peter, "The Case Against the Strategic Petroleum Reserve," *Policy Analysis*, No. 555, November 21, 2005.

[12] Bureau of Ocean Management, Regulation and Enforcement. *http://www.mrm* .boemre.gov/AssetManagement/ default.htm

[13] During the heating oil season, DOE updates and posts a weekly table that shows the various inputs that go into the calculation to determine the current differential, http://www.fe.doe.gov/programs/reserves/heatingoil/ Sales_Basis_0506.html.

[14] Personal communication with David Johnson, DOE Deputy Assistant Secretary for Petroleum Reserves, March 7, 2011.

[15] On March 1, 2011, Senators Jack Reed and Sheldon Whitehouse urged President Obama to sell the oil from the Bayou Choctaw cavern rather than move it to new cavern as proposed by DOE.

[16] These were sales ordered by Congress as deficit-reduction measures. For a chronology of these sales, see http://www.fe.doe.gov/programs/reserves/spr/spr-drawdown.html.

[17] Oil Daily, October 30, 2007. *Crude Continues Its Rally as Storm Hits Mexican Crude Exports*: p. 3.

[18] CRS Report R41478, *The U.S. Oil Refining Industry: Background in Changing Markets and Fuel Policies*, by Anthony Andrews, Robert Pirog, and Molly F. Sherlock.

[19] CRS Report R41478, *The U.S. Oil Refining Industry: Background in Changing Markets and Fuel Policies*, by Anthony Andrews, Robert Pirog, and Molly F. Sherlock.

[20] U.S. Government Accountability Office, *Strategic Petroleum Reserve - Options to Improve the Cost-Effectiveness of Filling the Reserve*, GAO-08-512T, February 26, 2008, p. 5, http://www.gao.gov/new.items/d08521t.pdf.

[21] See Government Accountability Office. *Options for Improving the Cost-Effectiveness of Filling the Reserve*. February 2008. GAO-08-521T, *http://www.gao.gov/new.items/*

d08521t.pdf. An additional benefit to acquiring heavier grades of crude is that it is less expensive.

[22] Use of refining capacity has been running at generally observed averages, taking into account seasonal maintenance and other events that will take refinery units offline temporarily. See Table 2 at http://www.eia.doe.gov/oil_gas/ petroleum/data_publications/weekly_petroleum _status_report/wpsr.html.

[23] See testimony of Rusco, Frank. Director, Natural Resources and Environment, Government Accountability Office, appearing in U.S. Congress. Committee on Energy and Natural Resources. Strategic Petroleum Reserve. S.Hrg. 111-67. May 9, 2009, p. 8-17.

INDEX

C

D

E

F

G

H

N

O

P

Q

R

S

T

V

W

Y